FIRST
IMPRESSIONS:
SHOPFRONT
DESIGN IDEAS
商业店面设计

（意）斯特凡诺·陶迪利诺／编
张晨／译

辽宁科学技术出版社

CONTENTS 目录

- 004 FOREWORD
  前言

- 006 PART/1 Introduction
  第 ❶ 篇 内容介绍

- 012 PART/2 Parts of a Shopfront
  第 ❷ 篇 店面的构成

  - 014 2.1 Fascia
    横带
  - 015 2.2 Cornice
    挑檐
  - 015 2.3 Window
    橱窗
  - 016 2.4 Entrance
    正门
  - 016 2.5 Polasters and Stall Riser
    壁柱和竖板

- 018 PART/3 Planning for Shopfront Design
  第 ❸ 篇 店面设计规划

  - 020 3.1 Key Considerations for Planning
    店面规划的主要考虑因素
  - 020 3.2 Planning Permissions and Consent
    规划许可
  - 022 3.3 Plans Required
    规划书要求

- 024 PART/4 Renovation of Traditional Shopfront
  第 ❹ 篇 传统店面的翻新

  - 028 4.1 History of Shopfronts
    店面的历史
  - 030 4.2 Renovation Principles
    店面改造原则

- 036 PART/5 New Shopfront Design Principles
  第 ❺ 篇 新店面设计法则

  - 039 5.1 Unity between Shopfront and Building
    店面与所在建筑的统一性
  - 042 5.2 Control to Scale, Height and Proportions
    对店面大小、比例和高度的把握
  - 044 5.3 Colour Scheme
    配色
  - 045 5.4 Lighting
    照明
  - 047 5.5 Choice of Materials
    建材的选择

050   PART/6   Design Approach of Architectural Elements
第❻篇　店面建筑元素的设计方法

- 052　6.1 Fascia and Signage Design
  店面横带和标识设计
- 061　6.2 Stall Riser Design
  竖板设计
- 062　6.3 Window and Glazing Design
  橱窗设计
- 062　6.4 Door and Access Design
  入口与通道设计
- 065　6.5 Blind and Conopy Design
  遮盖和遮阳篷设计
- 067　6.6 Display on Highway and Footpath
  放置在公路和人行道上的宣传展示

068　PART/7   Security
第❼篇　安保装置

- 071　7.1 External Shutters
  外部护窗板
- 071　7.2 Internal Shutters and Grilles
  内部护窗板和格栅
- 072　7.3 External Grilles
  外部格栅
- 072　7.4 Laminated Glass
  夹层安全玻璃
- 073　7.5 Other Measures
  其他安保措施

074   PART/8   Works
第❽篇　设计案例

- 076　8.1 Fashion
  服装店
- 178　8.2 Beauty
  美妆店
- 190　8.3 Catering
  餐饮店
- 246　8.4 Grocery
  食品店
- 280　8.5 Boutique
  精品店
- 306　8.6 Furniture
  家居店

318　INDEX
索引

CONTENTS

FOREWORD 前言

World Shopfront Design: DNA of Brand Building

Shopfront - as we say it in French the 'façade', is the face of the shop. It is obvious to everyone the importance of the shopfront - it gives the first impression of the shop to people. It is the entryway of the exterior of the shop towards the interiors. It is natural for architects and brands to spend huge amount of energy to conceive the best-designed shopfront to attract customers.

A shopfront works as the advertisement for the brand, its first function is to attract attention for the business and its merchandise. Trajan's Market in the heart of Rome, is thought to be the world's oldest shopping mall built almost over 2000 years ago. This multiple-level structure is ancient example of how shops were built with their façade. At this stage, there was still no separation of the structure and the wall. The façade was an elegant travertine frame. The external wall was painted what people could find inside. The external skin of the building, although many architects have spent great length to enhance its esthetics, remained the core structure to support the building. As time goes by, the evolution of shopfront design has been largely in-lined with the development in architecture. Since the 19th century, the availability of architectural cast iron has become more and more abundant. Architects and contractors have been able to experiment with iron columns and beams as the foundation of the buildings. At the same time, the development in glass-making technology has enabled the manufacture of large panel of glasses at a relatively lower cost. These two technological advancements together have gradually resulted in the shopfront as we know today: the structural elements are supporting large area of glass behind which the merchandise is displayed.

As the shopfront is usually built at the street level of the boutique, traditionally there are two ways to handle the shopfront image in relation to its surrounding. Firstly, many

世界商店门面设计：品牌建设的精粹

商店门面（在法语中称为"façade"）是一家商店的"面子工程"。商店门面的重要性对每个人来说都显而易见——它让商店给人们留下第一印象。它是从商店外部进入内部的通道。自然，建筑设计师和品牌投入大量精力，为吸引消费者设计出最为完善的商店门面。

商店门面宛如品牌广告，它的首要功能是为企业及其商品吸引关注。位于罗马市中心的Trajan's Market被誉为全世界最古老的购物中心，建造于大约2,000多年之前。这栋多层结构的建筑物就是商店如何构建门面的历史典范。当时，结构和墙身仍然连结在一起。其外立面运用了优雅的石灰华框架结构，外墙所采用的饰面与内部空间的装饰相同。尽管许多建筑师投入大量精力来增强外立面的美观度，但它依然是支持整栋建筑物的主要结构。随着时光的变迁，商店门面设计也随着建筑学的发展而逐步演进。自19世纪起，市场中的建筑铸铁产量变得日益丰盛。建筑师和承建商们得以尝试运用铸铁柱子和横梁来作为建筑物的基座。与此同时，玻璃制造技术的进步也让大块玻璃的制作相对不再那么昂贵。这两项技术进步融为一体，逐渐形成了我们今天所熟知的商店门面：结构元素支撑着大面积的玻璃，而在玻璃背后，则陈列着精美商品。

由于门面通常位于精品店的街面楼层，通常可通过两种方式处理商店门面的形象与周围环境的关系。首先，许多品牌营造出与周边环境形成鲜明对比的门面形象，令门面成为大胆醒目的关注点，突出品牌的存在，并向消费者传达着有力的信息。这种现象通常

brands create a shopfront image that is in big contrast with the surrounding environment, making the shopfront a bold and prominent voice to highlight the existence of the brand, sending a strong message to customers. This phenomenon is usually seen in brands that plan to carry a more energetic image towards the market. Secondly, other brands try to be more subtle in conceiving a shopfront more homogeneously fused with the surrounding environment. There is a big respect to the history and culture of the local area. Either way, the difference in these approaches is the result of the strategies of the brands, whether it aims to create a high-profile image or if it plans to be more low key. The balance between the boldness and the homogeneity has always been a delicate subject for architects and designers to explore, making façade design one of the most fascinating themes of architectural discussion of all time.

A shopfront sets the tone of the design of the brands. It shows the core values of the business. As the market is becoming more and more competitive, each brand strives to build up its unique image to create a distinctive soul - the DNA that cannot be replicated. The shopfront is a manifestation of the DNA in a nut shell. It is the biggest 'business card' to give away for free. It offers a glimpse of the merchandise inside. It is a teaser - customers are hypnotized as soon as they see the shopfront and become immediately tempted to go inside the store. It is the magic spell that calls for people's attention. As a result, façade design is essential to the brand building process.

As always, history is a mirror for us to learn from today. We need an adaptation to an always more demanding and sophisticated market but at the same time, it also gives the perfect opportunity for designers to deliver the best.

Stefano Tordiglione
Creative Director
Stefano Tordiglione Design Ltd

出现在希望向市场传达更具活力形象的品牌中。其次，其他品牌则在商店门面设计中试图表现得更加低调含蓄，与周围环境和谐融合。这代表着对当地历史和文化的尊重。无论采用何种方式，其差异都源于品牌的不同策略，无论其目标是营造出高调形象，还是计划更为低调。大胆和和谐之间的平衡始终是一大微妙主题，值得建筑师和设计师们探索，也令商店门面设计始终是建筑学讨论中最令人着迷的主题之一。

商店门面为品牌设计奠定基调。它展现出企业的核心价值观。随着市场竞争日益激烈，每个品牌都致力于构建独特形象，营造出与众不同的品牌灵魂——即是无法复制的DNA。商店门面是对整体品牌DNA的表达。它是企业最大的"名片"，而且可以"免费派送"。它让人们能够隐约窥见商店内部的商品。它宛如"预告"——让消费者慢慢地被催眠，让他们一看到门面就立即渴望进入商店内部。它仿佛是魔咒，吸引着人们的注意力。因此，商店门面设计是品牌建设流程的关键。

"以史为鉴，可以知兴替。"我们需要对日益要求严格、品位升华的市场做出调整，但也应让设计师们有机会展现最完美的作品。

斯特凡诺·陶迪利诺（Stefano Tordiglione）
斯特凡诺·陶迪利诺设计公司（Stefano Tordiglione Design Ltd）
创意总监

FOREWORD

PART 1

第 ❶ 篇 内容介绍

INTRODUCTION

Figure1.1 Bulgari New York, New York, USA, designed by Studio Marco Piva, photo by Andrea Martiradonna
图 1.1 宝格丽珠宝纽约店，美国，纽约，马克·皮瓦工作室设计，安德鲁·马迪拉多那摄影

A shopfront projects the best possible image of the business. It needs to display goods effectively and attract customers. It is in the shop owner's interest to make sure that the shopfront is well designed and makes a positive contribution to the street. Attractive shopping streets that provide a pleasing shopping experience will lead to higher custom. (See figure 1.1)

Shopfronts are the main advertising method for retailers and are required to be eye-catching as well as conveying the type of business offered. It creates the first impression of the trade with potential customers. A good shopfront should add interest to the street scene, attract shoppers and encourage them to stay and spend. In order to maintain the character of retail areas, a concerted effort from all involved is required. If well designed, shopfronts can make a positive contribution to the character and trading success of an individual street or the whole shopping centre.

The principle purpose of a shopfront is the advertisement and display of goods and services provided inside the building. Good design will reinforce the shop's identity and its location in the street, but by reflecting the style of the whole building above street level, and that of its neighbours. A good design will treat the shopfront as an integral part of the whole building and street frontage without focussing exclusively on the retail outlet alone. (See figure 1.2)

(See figure 1.3, 1.4, 1.5) Here, a good shopfront should be:
- The proportions of the shopfront should harmonise with the main building;

店面设计直接反映一家公司的形象。充分展示商品的同时，有效地吸引消费者是店面设计追求的目标。商店所有者希望店面形象设计精巧，对整条街道的消费体验有积极的提升作用。愉快的消费体验、令人心驰神往的购物街往往能构成良性循环，吸引更多的顾客。（图 1.1）

店面形象是零售商的主要宣传手段，店面必须要醒目，并且能够迅速传达店内提供商品或服务的信息。它是店铺给潜在消费者留下的第一印象。成功的店面设计能够为城市街景增添趣味，吸引消费者驻足欣赏，并鼓励他们进店消费。为了保持零售区域的特点，所有参与方的一致努力是必不可少的。优秀的店面设计可以为独立购物街或者整个购物中心的特色塑造和成功经营贡献一臂之力。

店面设计的主要原则是实现广告宣传作用，并且能够充分展示店内经营产品或服务内容。优秀的设计方案能够使店铺的品牌形象得到强化，在其所在区域树立地位，并在超越街道及其他店铺的层面上体现整个店铺的风格和个性。这样的设计将店面和整个建筑、整条街道当做一个有机整体，而不仅仅局限于零售网点本身。（图 1.2）

（图 1.3，图 1.4，图 1.5）成功的店面设计一般具备以下几个特点：
- 店面比例与所在建筑协调；

Figure1.2 Guru Palermo, Palermo, Italy, Designed By Duccio Grassi Architects, Photo by Duccio Grassi Architects
图 1.2 巴勒莫大师品牌旗舰店，意大利，巴勒莫，杜乔·格拉西建筑师事务所设计，杜乔·格拉西建筑师事务所摄影

Figure1.3 Freitag Store Tokyo, Tokyo, Japan, designed by Torafu Architects, photo by Sebastian Mayer
Figure1.4 Bolo ao Forno!, SaoPaulo, Brazil, designed by Luciana Carvalho and Renato Diniz, photo by Cezar Kirizawa
Figure1.5 Global Style Tokyo, Tokyo, Japan, designed by Process5 Design, Photo By Process5 Design

图 1.3 弗赖塔格东京店，日本，东京，Torafu建筑设计公司设计，塞巴斯蒂安·梅耶摄影
图 1.4 Bolo ao Forno 甜品店，巴西，圣保罗，卢恰娜·卡瓦略，雷纳托·迪尼兹设计，切扎尔·克里扎瓦摄影
图 1.5 Global Style 东京店，日本，东京，PROCESS5 DESIGN 设计公司设计，PROCESS5 DESIGN 设计公司摄影

PART 1 Introduction

Figure1.6 Skandia Banken, Sweden, designed by BVD, photo by Åke E:son Lindman
图 1.6 斯堪的亚银行，瑞典，BVD 设计公司设计，阿克·艾森·林德曼摄影

- Materials should reflect the existing range on the original building;
- The shopfront should not be treated separately from the upper levels;
- It should add interest and attract custom;
- It should avoid standardisation, reflecting the diversity of a street scene.

As shops change hands or need refitting, there can be significant pressure to update and modify shopfronts. Without sensitive design, successive changes may fail to project a positive image for the retail unit or the street and can undermine a place's appearance and attraction. Common problems include: alterations of a hasty or temporary nature, clashing or dominantly coloured or over sized components, badly maintained units, or design that pays little regard to the building, street and area within which it is located.

・使用材料反映原有建筑的材质特点
・店面与所处环境协调，融洽
・店面应吸引消费者，刺激消费行为的发生
・避免标准化设计，体现所在街道街景的丰富性和多样性

当店铺易主或单纯的进行改装时，店面的改造工程可能会面临重重困难。一旦缺乏合理的安排和改造，后续的改建工程可能无法为店铺以及整个商业街塑造积极的品牌形象，有的甚至还会投射负面的影响。这一环节常见的问题有：仓促决定或临时的变动，与其他元素冲突、色彩过浓、尺寸超标的部件，或者与所在建筑、街道和区域不协调的设计。

Here, some practices will normally cause the problems above:
- Any forms of advertisement that detract from the appearance of the host building or the special architectural interest of the County's conservation areas.
- Internally illuminated box fascias, individually illuminated letters, halo-lit perspective letters, fluorescent lighting on channels and illuminated projecting box signs.
- Materials such as Perspex, acrylic sheeting, uncoated aluminium and glossy plastics.
- Further use of swan neck lights and spotlights in conservation areas.
- The use of shiny or garish materials for fascias, box signs or hanging signs.
- The introduction of Dutch blinds.
- To minimise visual damage within a conservation area the council will encourage the use of internal grilles mounted behind windows.
- Where internal grilles are not possible, some types of external grille, such as removable grilles, may be acceptable. In some instances, a lightweight see-through mesh grille might be considered if it is incorporated behind the fascia and painted to match the shopfront.
- Security systems which incorporate external box housings or solid shutters.

This book provides guidance to improve the standard of shopfront design and advertisements throughout the world. The purpose is to encourage greater care to promote high quality design standards in order to create settings in which retailers can establish and develop successful businesses. The book is not intended to be overly prescriptive or stifle modern innovate designs. It is to provide an understanding of the design of shopfronts and advertisements that shopkeeper and local government will support when reaching a decision on any planning application or application for advertisement consent. (See figure 1.6)

The objectives of the book are to provide a consistent and integrated approach towards the design of shopfronts, to assist designers to achieve high quality shopfronts that are accessible to all through inclusive design, and to ensure that the design of shopfronts contributes positively to the vitality of the relevant areas daytime and evening economy without detriment to safety and security.

In this book the term 'shop' is defined as any commercial premises having a fascia sign or display window, including non-retail premises such as banks, restaurants, takeaways, estate agents and other businesses in a shopping area.

This book supplies with advice and specific design principles and approaches. Proposals for new or altered shopfronts and advertisements are likely to meet the requirements. The principles and approaches in the book are relevant where works to a new shopfront or the installation of a shopfront are proposed within a conservation area or in relation to a nationally or locally listed building. These principles and approaches seek to strike an appropriate balance between the need for development and the conservation of heritage assets.

The book has been arranged into six main sections:
1. Parts of a shopfront
2. The planning of shopfront design
3. The renovation of traditional shopfront
4. New shopfront design principles
5. Design approach of architectural elements
6. Security design

下面列举的是可能导致以上问题的原因：
- 对店铺所在建筑主体或有保护价值的特别建筑构成干扰的任何形式的广告
- 内部照明箱、有独立照明装置的字母、卤素字母、有轨道荧光灯和灯箱等照明因素
- 有机玻璃、亚克力板、无涂层铝材和亮面塑料等材料的使用
- 保护区内鹅颈灯和聚光灯的不恰当使用
- 广告牌、灯箱或悬挂式标识等使用反光或刺眼材料
- 百叶窗的使用
- 为了最大程度地减少保护区内的视觉破坏，有关当局鼓励在窗户内侧使用内部格栅
- 无法使用内部格栅的地方，一些形式的外部格栅，如可拆卸式格栅也是可以接受的。一些情况下，可以在标识后方使用与店面为外观同色的轻质网状格栅
- 连接外部接线盒或固定挡板的安保系统

本书适用于全球范围内的店面设计，为实现更高水平的店面和广告设计提供了切实有效的参考和指导，鼓励从事该专业的人员要追求更高的设计标准，设计出有助于零售商建立并发展生意的场所。本书并不关注机械教条式的内容，而是着眼于分析理解店面设计的精华，在店主的需求、品牌的形象和有关部门所制定的规定之间达到平衡。（图 1.6）

本书希望为店面设计师提供综合全面的参考和支持，确保店面设计方案对所在区域增添积极的活力与影响，在保证安全的前提下推动消费和经济。

本书中"店面"所指的范围涵盖任何配有招牌和橱窗的商业处所，也包括银行、餐馆、外卖店、房产中介和购物区内常见的其他非传统商业场所。

本书收录了针对设计过程的建议以及具体的设计法则，在新店设计或旧店改造工程中具备很好的实用性和参考价值。这些设计法则和方法同样适用于受保护区域或建筑内的店面建设和改造。掌握这些设计原则和方法有助于实现商业发展与传统保护之间的平衡。

本书分为 6 个主要部分：
1. 店面的组成
2. 店面设计规划
3. 传统店面的翻新
4. 新店面设计法则
5. 店面建筑元素的设计方法
6. 安保装置

PART 2

PARTS

第 ❷ 篇　店面的构成

OF A SHOPFRONT

Figure 2.1 Sketch of shopfront
Figure 2.2 1010 Tsimshatsui. Flagship Store, Hong Kong Tsim Sha Tsui, China, designed by Clifton Leung, photo by Shia Sai Pui
图 2.1 店面草图
图 2.2 1010 尖沙咀旗舰店，中国香港，尖沙咀，梁显智设计，佘世培摄影

1. Cornice
2. Blind Box
3. Frieze
4. Pilaster
5. Window
6. Entrance
7. Stall riser
8. Mullion
9. Transom
10. Architrave
11. Console Bracket
12. Fascia

1. 挑檐
2. 百叶窗匣
3. 装饰带
4. 壁柱
5. 橱窗
6. 正门
7. 竖板
8. 窗棂
9. 气窗
10. 框缘
11. 托座
12. 横带

Although shopfront design should be seen as a whole, it is made up of component parts, each of which has its own visual and practical function. These features define the style, and help integrate it into the rest of the building.

Shopfronts are composed of functional parts which together form a complete visual composition. Each part has a specific role. These apply equally to any period of construction, not just shopfronts on historic buildings or in conservation areas. They are a sound basis for designing shopfronts, including modern design. The key parts of a traditional shopfront are: cornice, fascia, windows, entrance, pilasters and stall riser. These parts effectively enclose the shop window and entrance in the manner of a picture frame. They direct the eye to the entrance and provide a solid 'base' for the building above. The pilaster identifies the vertical division between shop-fronts; the fascia provides advertising space and the stall riser gives protection. (See figure2.1)

2.1 Fascia

The fascia board is located across the shop between the console brackets at the top of the pilasters. This is the area used to display the shop name and was traditionally angled towards the street to be easily read.

The fascia is probably the most important and noticeable element of a shopfront. It has the potential to have a major impact on the quality of the street scene. It should be seen as an integral part of the shop-front, and not just as a form of advertisement. It needs to be

店面设计尽管应被当做整体处理，却是由不同的部分组成的，每个部分有各自的观赏和实用功能。这些元素构成店面的整体风格，并帮助店面融入所在建筑环境。

店面的功能元素组成一个完整的视觉效果。每个部分都有其特定的功能，并作用在施工的每个阶段，对于普通建筑和需要保护的历史建筑同等适用。也正是这些元素构成了店面设计的坚实基础。传统意义上的店面功能结构包括挑檐、横带、窗户、正门、方壁柱和竖板。这些元素像相框一样，将窗户和门纳入一个较为紧密的视觉范围，有效地将人们的目光引向店铺入口，也为店面上方的整个建筑设下一个稳固的"基础"。方壁柱在竖直方向上区分相邻的店面，横带适合摆放广告牌，竖板则起到一定的保护作用。（图 2.1）

2.1 横带

横带是店面中壁柱顶端、装饰性支架之间的横向楣板，通常用于展示店铺的名称。横带朝向街道方向，能够迅速传达店内提供商品或服务的信息，起到宣传的重要作用。

横带可能是店面中最明显也是最重要的一个部分，能够对整个街道景观起到较为直接的影响。设计师应该将横带视为店面的一个必须元素，而不仅是一种宣传手段。选择适合的字体、风格以及

appropriate in character, style and proportion to the building.

2.2 Cornice

Above is a cornice which provides a distinctive horizontal divide between the shop and the upper floors.

The Cornice is both a decorative and functional feature of the shopfront. In terms of decoration it forms a conclusive termination to the top of the fascia and thus the shopfront as a whole. Functionally, it projects forward of the shopfront throwing water clear of the fascia preventing water ingress and reducing the incidence of rot. It is also common to find a roller blind incorporated within the cornice.

A projecting moulded cornice protects the fascia and shop below from rainwater runoff. Decorative carved console brackets form 'bookends' to the fascia between the cornice and pilaster. They help frame the fascia and add vertical rhythm to the shopfront. It also throws rainwater clear of the shopfront and prevents decay. A structural or applied cornice projection is required as part of nearly every shopfront design.　(See figure2.2)

2.3 Windows

Windows are subdivided by transoms and mullions to form horizontal and vertical divisions. They form a large visual element in the shopfrontage and are used to display goods and

与整个建筑协调的比例是横带设计中需要注意的几个因素。

2.2 挑檐

上图的挑檐设计在水平方向上起到了分隔店面与上层空间的作用。

挑檐具有装饰性，同时也是店面中一个不可或缺的功能性结构。从装饰性层面出发，挑檐与横带的上端边缘衔接，构成和谐统一的店面外观。功能方面，挑檐将雨水导离横带方向，保护横带免受雨水浸泡，减少腐蚀的发生。卷帘与挑檐搭配使用也是十分常见的。

挑檐将雨水导离横带和其他店面结构，对店面起到保护的作用。装饰支架对挑檐和壁柱之间的横带结构起到"书挡"一样的作用，并且对横带也起到一定的限定作用，还为店面的竖直方向增添了韵律感。几乎所有的店面设计都会包含结构或实用意义上的挑檐结构。（图2.2）

2.3 橱窗

橱窗在店面中可以细分为横窗和竖窗，也是店面的大型构成元素，它通常用来展示最前沿的商品，起到吸引消费者眼球的功能。在

Figure 2.3 Salon1, Bielefeld, Germany, designed by Kiss Miklos ,photo by Sören Münzer, Fanni Kovács & kissmiklos
Figure 2.4 Farm – Niterói, Niterói, Rio de Janeiro, Brazil, designed by be.bo. , photo by Marcos Bravo
图 2.3 沙龙一号店，德国，比勒费尔德，葛斯·米罗斯设计，瑟伦·闵采尔与范尼·克瓦斯和葛斯·米罗斯设计工作室摄影
图 2.4 FARM 精品服饰店，巴西，里约热内卢，尼泰罗伊，be.bo. 设计公司设计，马科斯·布拉沃摄影

attract customers. Vertical divisions often reflect the vertical division of the upper floors. The cill supports the windows and, like the stall riser, provides protection.

The size and style of windows, glazing bars, mullions and transoms should be in scale and proportion with the rest of the shopfront and the building as a whole. The number, location and dividing up of any glazed areas must relate to the upper floors and any adjoining buildings.

2.4 Entrance

The entrance is typically centrally located and from the late 19th century often became recessed to give visual interest, shelter and maximum window display. Entrance gives an important first impression and can have a significant impact on the appearance of the building. The design and positioning of the door should reflect the character of the whole building. (See figure2.3, 2.4)

2.5 Pilasters and Stall Riser

Pilasters are shallow piers or columns that project slightly from the wall on each side of the shopfront, and give vertical framing and visual support to the fascia and upper floors and form a type of picture frame, which are flat or decorated columns which define the width of the shopfront and enclose the window frame. A pilaster establishes a visual separation between neighbouring properties. They are a traditional building feature designed with a base and capital. Above the pilasters are projecting heads known as consoles. The base of

橱窗的下方通常设置窗台，窗台对橱窗起到支撑作用，并且与竖板一样，能够提供保护。

橱窗、玻璃隔条、横窗和竖窗的尺寸以及他们的设计风格应与整个店面的面积、整个店面的风格以及所在建筑的风格相匹配。玻璃面的数量、位置和分隔方式应与店面上方楼体以及相邻建筑的分布模式相匹配。

2.4 正门

店铺正门通常位于店面的中心位置。19 世纪后期开始，西方店面的正门设计以凹进的形式为主，这样不仅具有视觉吸引力，形成一个遮风挡雨的小空间，而且方便进行最大程度的窗口展示。正门可以给人留下深刻的第一印象，对整个建筑的外观也有着直接的影响。店面正门的设计和定位与整个建筑的风格有着紧密的联系。（图 2.3，图 2.4）

2.5 壁柱和竖板

壁柱是店面两侧，在竖直方向上构成视觉框架，并略微突出墙面的间隔物。壁柱能够对横带和上层建筑起到视觉支撑作用。无论光滑平坦还是有图案装饰，壁柱对店面的横向范围有明确的指示作用，并对窗口起到聚拢统一的效果。同时，壁柱还可以分隔相邻店面，属于一种传统的建筑元素。壁柱上方通常与支架连接，

Figure2.5 Chocolat Milano, Parma, Italy, designed by Blast Architects, photo by Pietro Savorelli
图 2.5 米兰巧克力店，意大利，帕尔马，布拉斯特建筑师事务所设计，彼得罗·萨沃雷利摄影

the pilaster usually terminates in a plinth block and the head in a plain or decorated console bracket, or corbel, which supports any overhanging fascia.

Pilasters and consoles vary from being very elaborate and highly decorated to being relatively plain but they usually have some moulding or surface decoration. Where traditional pilasters and console details exist they should be retained. If new ones are introduced they should be designed to reflect the level of detail in other elements of the shop-front and constructed of an appropriate material.

The stall riser or stallboard as it was formerly called, is a long established shopfront feature originally housing boards or stalls, which hinged out over the pavement and carried goods for display. The stall riser is the area of the shopfront below the display window. Stall risers are now more commonly incorporated as a feature giving physical and visual support to the shopfront providing it with balanced proportions.

This protects the bottom from kicks and knocks and screens unattractive floor areas from public view. Visually it completes the frame enclosing the display space and it also helps give the impression that the shopfront is anchored to the ground. Fine attention to detail in the design of the stall riser can contribute greatly to the overall design of the shopfront. It provides a protective area between the shop window and the street level. It also adds a sense of security. It is often constructed of stone, brick, render or paneled timber. Appropriate heights will usually be between 450mm and 700mm. It also helps to provide a horizontal link to adjoining buildings. (See figure2.5)

支撑各种形式的横带。壁柱的基部以底座的形式为主。

壁柱和装饰支架样式繁多，可以精美复杂，也可以简洁平淡，但通常都是倒模制成，表面有一定的装饰。店面改造中，原有壁柱和装饰支架如有图案，则应对其进行保留。如果需要增加新的壁柱，则应选择装饰精细程度与店面其他元素相互匹配的材料。

竖板也称铺面栏板，是一种传统的店面结构元素，最初用来存放宣传板和小货摊。通常位于路面中央，进行商品展示。竖板是店面橱窗下方的区域，如今已经发展为店面设计中兼具实用和观赏功能的构成元素，对协调整个店面的比例有重要作用。

从实用性来看，竖板能够抵挡来自外界的踢踩和敲击，遮挡不甚美观的地板区域，呈现相对完美的店面形象。竖板与其他元素连接，构成完整的展示区域，也给人一种踏实稳固的感觉，这是它的观赏性功能。竖板区域精致的细节设计能够极大地提升店面的整体设计质感。由于竖板在店面橱窗与街道之间形成保护，也能够提升安全感。石材、砖、粉刷、板材是常用的竖板材料。通常来说 450～700 毫米是适当的高度范围。最后，在水平方向上，竖板也能与相邻建筑形成视觉上的连接。（图 2.5）

PART / 3

第 ❸ 篇 店面设计规划

PLANNING FOR
STOREFRONT DESIGN

Table 3.1 Key Considerations
表格 3.1 主要考虑因素

3.1 Key Considerations for Planning

It is always advisable to undertake early discussions with the Planning Authority in order to obtain their initial views on the proposals and ascertain what permissions are required. By carrying out pre-application discussions for any proposed scheme, detailed advice can be given in order to ensure that any formal submission accords with the objectives of the Local Plan policies and planning document.

Prior to making a planning application it would be advisable to consider the following points: (See table 3.1)

3.2 Planning Permissions and Consent

Most alterations to shopfronts will require approval under the Planning Acts, Advertisement Regulations and Building Regulations. More than one type of consent may be required. Before making any alterations, developers are advised to contact Planning Services to check if consent is needed.

Planning permission will be required for any alterations that materially affect the external appearance of the shopfront, such as replacement of the shopfront or frame, changes to the fascia, the installation of external security shutters and grilles, the installation of a canopy/blind or awning, or illuminated signs. Permission is also needed for changes to the materials used. Works that do not materially affect the appearance of the shopfront, such as repainting and maintenance, do not require planning permission.

3.1 店面规划的主要考虑因素

建议设计师在确定了店面设计的基本方案后，应尽早与市政规划部门以及店主进行进一步的沟通，以获得初步的修改意见，并且了解需要申请的许可类型。通过初期调查讨论获得详尽的申请建议有利于提高效率，制定出符合当地规划改造规章政策的工程方案。

在制定规划申请之前，建议设计师首先考虑以下几点主要因素。（表格 3.1）

3.2 规划许可

大部分店面改造工程都需要符合当地的相关规划法规、广告条例和建筑法规，并且可能需要获得多方批准。在设计师对店面改造开始之前，开发商应该与市政规划部门确认是否需要获得相关许可。

例如对店面进行翻新、横带的修整、店面外部安装安全帘、安装格栅、安装遮阳棚或灯箱广告等任何对店面外观产生实质性改变的工程都应该向上级部门申请规划许可。如要改变该店面所在建筑的材料则同样需要提交申请许可。重新粉刷以及对店名进行日常维护等对店面外观不构成实质性改变的工程则无需申请规划许可。

	Key Considerations: 主要考虑因素
About the building 建筑因素	Is the building listed or in a conservation area? 店面所在建筑是否为受保护建筑？所在区域是否为受保护区域？ What is its history, is the existing shopfront original, even if in part, or is it modern? 建筑有怎样的历史？现有店面是否保留了最初的全部或部分设计？ Are these any original features under the box fascia or cladding? 灯箱和骨架外墙下方是否保留了店面之前的设计？ Are there any old photographs which show the historic shopfront? 是否可以借助老照片参考店面以前的外观？
About the design 设计因素	Does the proposal take account of the principles of good shopfront design? 计划书是否涵盖了成功店面的设计准则？ Does it respect the design and scale of the building? 店面规划是否与建筑本身的设计和规划相冲突？ Does it take account of vertical and horizontal emphasis on the street scene? 设计是否考虑到竖直和水平方向上的街道景观？ Are the materials used appropriate to the building and its location? (See figure 3.1) 店面计划使用的建材是否与建筑及周边环境相符？（如图 3.1） Is the detailing appropriate for the building and its location? (See figure 3.3) 店面的装饰细节是否与建筑及周边环境相符？（如图 3.3） Is the colour proposed appropriate for the building and its location? (See figure 3.2) 店面的配色是否与建筑及周边环境相符？（如图 3.2） Has the scheme fully considered sustainability issues? 设计是否全面地考量了环保的问题？ Has access for all been achieved? 设计是否考虑了老弱病残人群的使用需求？
About the security 安全因素	Have security measures been considered as an integral part of the overall design of the shopfront? 安保措施是否作为整个店面设计的一部分被纳入考虑？ What impact do the security measure have on the visual appearance of the building and street scene? 安保措施对整个建筑和街道景观有怎样的视觉影响？

Figuro 3.1 Bulgari, Paris, France, designed by Studio Marco Piva, photo by Andrea Martiradonna

图 3.1 宝格丽珠宝巴黎店，法国，巴黎，马克·皮瓦工作室设计，安德鲁·马迪拉多那摄影

Any alterations to a listed building require listed building consent if the works affect the character or appearance of the building. This can include small changes to features such as doors, decorative details, and fire and burglar alarms. It is always advisable to contact the local conservation team for advice on works to a listed building.

Consent is required for most advertisement works, such as installing a new fascia or projecting sign, or changing the materials or colour of a sign. Most illuminated signs require advertisement consent. In conservation areas and on listed buildings all illuminated signs require consent. The regulations can be complex and it is advisable to seek advice from the local relevant department. (See figure3.1, 3.2, 3.3)

In addition to planning and advertisement consent, certain works to shopfronts need to comply with building regulations legislation. For example, if work involves structural alterations, alterations to access and approach, or if there are implications for fire escape.

3.3 Plans Required

The following drawings will be required for planning and/or Listed Building Consent applications:
- Site location plan: to a scale not less then 1:2500
- Block plans: to a scale not less than 1:500
- Building plans: to a scale not less than1:50; to show elevations as existing and as proposed
- Detail plans: to a scale not less than1:50; to show all new doors, shopfronts, mouldings and joinery details and other decorative detail.

在受保护建筑周边进行的工程如果对该建筑的风格或外观有任何改动，则应该申请受保护建筑改造许可。对门、装饰性细节、火警和防盗报警系统等较小的改动都应受到重视。在受保护建筑的店面改造工程中与相关机构始终保持沟通是十分必要的。

大部分的广告设计，如安装新的横带或突出标识，将原有标识改用其他建材或配色等，都需要获得许可方能实施。大多数照明标识也需要申请广告许可。在受保护区域以及受保护建筑周边，所有形式的照明标识都需要获得许可。相关法规较为复杂，建议设计师向当地有关部门寻求建议。（图3.1，图3.2，图3.3）

除了规划和广告需要申请相应许可，店面设计工作还应符合当地建筑规范法则。例如，涉及建筑结构调整，建筑出入口改动和火警逃生路线的工程。

3.3 规划书要求

以下是工程涉及受保护建筑时，设计师需要遵循的标准：
- 总平面图：比例尺不小于1:2500
- 楼宇平面图：比例尺不小于1:500
- 建筑设计图：比例尺不小于1:50；应显示原有和设计中的立面图
- 细部图：比例尺不小于1:50；显示所有的新门、店面、装饰细节设计

Figure3.2 Bulgari, New York, USA, designed by Studio Marco Piva, photo by Andrea Martiradonna
Figure3.3 Bulgari, Rome, Italy, designed by Studio Marco Piva, photo by Andrea Martiradonna

图 3.2 宝格丽珠宝纽约店，美国，纽约，马克·皮瓦工作室设计，安德鲁·马迪拉多那摄影
图 3.3 宝格丽珠宝意大利店，意大利，罗马，马克·皮瓦工作室设计，安德鲁·马迪拉多那摄影

PART, 4

TH
TRADITI

第 ❹ 篇 传统店面的翻新

THE RENOVATION OF TRADITIONAL SHOPFRONT

There are many instances where an original shopfront will remain, hidden underneath later additions. For example, older fascias often lie beneath modern box fascias and pilasters and corbels were often boxed in as fashions changed. Where there is evidence of original details surviving, it may be desirable to retain and restore or refurbish any such features.

In Britain, the city council will encourage owners first to repair original shopfronts; second to repair or reestablish the traditional architectural frame of a shopfront; and third to propose well proportioned, high quality, modern design as a third alternative. (See figure 4.1)

A shopfront may provide a reference to an areas traditions, or form part of a historic feature that is nationally or locally significant. A sensitive approach to design is likely to be required in instances where they:
· Form part of a Listed Building;
· Lie within a Conservation Area;
· Contain, or are important features in their own right;
· Are representative of a local or wider style.

Where an existing shopfront is sympathetic to the building or of historic interest it should be refurbished and repaired rather than replaced. The detail, modelling and decoration of older shopfronts is particularly valuable in the street scene and their retention should always be considered. Where early shopfronts survive special care is needed to ensure that they are preserved and restored in a sensitive manner with careful attention to detail. Where an original shopfront has been altered much of the architectural framework, such

在很多情况下，原有的建筑外立面设计会被后续的改造工程所隐藏、覆盖。例如，现代的灯箱往往会将老式的横带遮挡起来，壁柱和托臂则按照当下的风尚被包围起来。设计师在工程设计中如果遇到可能保留的原有店面装饰，建议对其进行保留或修复。

英国的市政厅会鼓励房主对老旧的店面优先进行修复，其次选择重建老旧店面的建筑结构，最后再选择进行重新设计的高质量现代店面建设。（图 4.1）

店面不仅可以反映当地的传统文化，也可能在当地乃至全国树立独特的历史地位。以下情况通常需要设计师采取谨慎的设计方案：
· 工程涉及受保护建筑
· 工程位于受保护区域中
· 工程包含或属于重要的景观
· 工程代表当地或更大范围内的某种典型建筑

当目标店面与所在建筑及相应历史价值联系紧密时，设计师应尽量选择翻新修整而不是全盘重建。原有店面的细节安排、设计模式和装饰风格都对街景有着独特的价值，设计师应考虑保留这些元素。如果决定对老旧店面进行保留，应在施工中给予特别注意，确保原有结构得到的精心与恰当的修复和保护。对原有店面进行改造时，大部分原有建筑结构会被保留。设计师

Figure4.1 Zlatarna Celje Jewellry Flagship Store, Maribor, Slovenia, designed by OFIS Architects, photo by Tomaz Gregoric, Jan Celeda, Giulio Margheri
Figure4.2 Como quieres que te quiera, Buenos Aires, Argentina, designed by The Wall Arquitectura y Diseño, photo by Julio Masri
图4.1 Zlatarna Celje珠宝旗舰店，斯洛文尼亚，马里博尔，OFIS建筑师事务所，塔马兹·格雷戈里克，让·谢利达，朱利奥·马格里摄影
图4.2 "如果我想要……"服装店，阿根廷，布宜诺斯艾利斯，The Wall设计公司，胡里奥·马斯里摄影

as pilasters or fascias boxed in and hidden by later work often survives, and these can be revealed, greatly enhancing the appearance of the shopfront and the character of the street. Many C19 and early C20 shopfronts are of high quality and are worthy of retention.

If the shopfront is within Conservation Areas, on a listed building or building of local interest, where there is evidence of the original shopfront on older buildings, the reinstatement of traditional designs is encouraged. Alterations to existing shopfronts and any new works undertaken should not conceal or remove original or traditional detailing. Wherever possible any works carried out to original shopfronts should endeavour to reinstate any traditional features lost over the course of time. (See figure 4.2)

The removal of a traditional shopfront that is part of a listed building or within a conservation area will not be permitted if it is appropriate to the building or is of architectural or historic interest in its own right.

Permission for the replacement of traditional shopfronts in Listed Buildings or in Conservation Areas will only be granted if the existing shopfront is inappropriate to the building or area and its replacement will be of high quality and improve the character of the building or area.

The replacement of original shopfronts will only be considered where it can be fully justified. Other shopfronts, which may not be original but are of a high standard of architectural quality, including modern and replacement shopfronts which are deemed to enhance the area should also be retained where possible.

如果能对那些被遮挡的壁柱或横带结构加以利用，一般可以使店面外观和街景风格获得极大的提升。19世纪和20世纪早期的许多店面设计水平一流，极具保留价值。

如果目标店面位于该城市受保护的区域之内，是受保护建筑的一部分，或是在当地具有特殊价值，则建议设计师采用重现传统店面设计的方案。对原有店面进行的改动或任何形式的附加设计应该在不遮挡原有的传统装饰的前提下进行。条件允许的情况下，所有的店面改造工程都应努力重现店面原有的风貌。（图4.2）

在受保护的建筑内或者受保护的区域内，如果传统店面与该建筑风格相符或者具有建筑、历史价值，则必须将其保留，不得完全拆除。

受保护建筑内或者受保护区内的传统店面，只有其风格与所在建筑或区域定位不符时，才会获得批准进行改造。取代它的新店面须具备较高设计水平，能够对整个建筑或区域的景观起到提升作用。

店面改造工程只有在绝对需要改造的情况下才会被政府批准。其他建筑结构处于较好状态的店面，无论风格现代与否，只要能对所在地区的城市街道形象起到一定的提升作用，都应当对其进行保留。

Then, a modern design can sometimes be successfully incorporated into traditional building façades where careful consideration is given to: the age, style and proportions of the building, materials, and craftsmanship.

In some locations a modern shopfront will be appropriate and new shopfronts of innovative design are encouraged. Good modern designs are often based on the re-interpretation of traditional forms. A design could be developed within the traditional architectural framework or within a new shop frame that re-interprets the proportions of adjacent shopfronts in a contemporary way. The surround should look capable of supporting the upper floors and the design should add visual interest to the street. (See figure 4.3)

4.1 History of Shopfronts

It is useful to understand a little about the history of shopfronts so that the appropriateness of designs for specific buildings can be more fully appreciated. It is important to try to understand what features of a shopfront make a positive contribution to the character of the unit or the area. The period when the original building or street was constructed can help to provide useful clues for future proposals.

Medieval
The idea of shopping as it is known today is a relatively recent development. For centuries, goods were spread out onto the street or displayed on a drop-down shutter that served as a counter during the day.

为了让现代风格的设计方案也可以很好地融入传统店面设计，设计师应仔细地考虑建筑年份、风格、结构比例、使用材料和建筑工艺。

在一些情况中，现代的店面方案会具有较高的适用性，设计师可以大胆进行创新设计。优秀的现代店面设计通常建立在重新解读原有传统店面设计的基础上。设计方案可以在传统建筑框架内展开，或在重新解读相邻店面比例的新店面框架内进行。后者通常具备现代设计风格。选择的包围物应该在视觉上足够支撑上层结构，整个设计也应为街景增加一定的视觉吸引力。（图4.3）

4.1 店面的历史

对于设计师来说，对店面的历史具备基本的了解有助于把握设计的尺度，针对具体的建筑做出更为恰当的判断。知晓怎样的店面元素能够对所在建筑单元和区域构成积极的作用，这十分重要。了解建筑或街道初建时的历史背景也利于获得线索并寻找设计灵感。

中世纪
现代化的购物概念在历史上出现的较晚。几个世纪以来，货物都被摆放在街上，进行展示和售卖。

Figure4.3 Ferrari Store, Madrid, Spain, designed by Iosa Ghini Associati, photo by courtesy of Ferrari and Nicola Schiaffino

图 4.3 法拉利旗舰店，西班牙，马德里，马西默·尤萨·基尼设计工作室设计，法拉利品牌，尼古拉·斯基亚菲诺摄影

C18

From the C18 onwards, the shopfront became an integral part of the design of the building. A typical form may have a simple frame to the windows comprising of vertical columns or pilasters supporting a horizontal component known as the entablature. The design is likely to be based on Greek or Roman proportions, made from timber and may have projecting bow windows with small leaded panes in Crown glass (blown glass). Display windows often took the form of square bays or bow windows. Window panes were small and detailing was often in the classical style, which was the architectural fashion of the time. Timber shutters are often found.

Late C18 to mid C19

The classical style became more pronounced. A typical form may follow the essence of the Georgian style but with increased ornament and functionality with additional components such as roller blinds. The shop window was framed by pilasters, which provided visual support for the top, or entablature. Projecting bays were now outlawed in most places to avoid obstructing the pavement. Larger planes of glass are likely. Materials other than timber start to be used more extensively.

Victorian

More emphasis was now given to the name of the shop and the fasçia was emphasised at the expense of the cornice. Sometimes the fasçia was tilted to accommodate a blind box. Console brackets appeared at either end of the fasçia. Decoration often became more exuberant and later on a variety of materials, such as bronze, cast iron and terra cotta were introduced. The invention of plate glass saw the appearance of larger window panes.　Early

18 世纪

从 18 世纪开始，店面逐渐被人们重视起来，成为建筑设计的一个组成部分。竖立的柱子或壁柱支撑被称为柱上楣构的水平结构，配合窗户就构成了典型的店面设计。当时的这种设计可能受到希腊或罗马审美比例的影响，通常以木材为主。使用皇冠（吹制）玻璃的突出的弓形铅条玻璃窗在当时也较为常见。橱窗中比较常见的是方形或弓形窗口，这种窗框比较小巧，装饰的细节部分以当时流行的古典建筑风格为主。木制百叶窗也开始被人们使用。

18 世纪末期 –19 世纪中期

这一时期，古典建筑风格更为明显。常见的店面设计遵从乔治风格的主要理念，增加的卷帘等元素使装饰性和功能性都得到了提升。店铺橱窗以壁柱为框，在视觉上起到支撑顶部结构的作用。此时大部分地区对突出的橱窗设计进行限制，以防阻碍通行。选用更大的玻璃已经逐步成为可能。木材以外的其他建筑材料开始被广泛地使用。

维多利亚时期

这一时期对店铺名称较为重视，横带也取代了挑檐成为店面设计中的一个重点。有时也把百叶窗匣放在横带的位置。装饰性支架通常位于横带一端。装饰细节愈发丰富多彩，并逐渐开始使用青铜、铸铁和赤土等材料。厚玻璃被发明后，大玻璃窗也在店面设计中多了起来。

Figure 4.4 Money Shop, Birmingham, UK, designed by Nicolas Tye Architects, photo by Nicolas Tye Architects
图 4.4 钱庄，英国，伯明翰，尼古拉斯·泰伊建筑师事务所设计，尼古拉斯·泰伊建筑师事务所摄影

C20

Generally, the established principles of shopfront design stayed the same but styles were often adapted to emphasise the type of shop. Easily cleaned glazed tiling, for instance, was fashionable for butchers or pubs.

Later C20

The 1960s and 70s saw a radical change in design philosophy and traditional design was no longer venerated. Over-large windows, dominant fasçias, cheap materials and disregard for existing buildings obliterated the character of many shopping streets.

4.2 Renovation Principles

Historically, ground floor shopfronts were designed as an integral part of the building and as such they related to and incorporated features of the building above.

Traditional shopfronts predominately developed around a set of common elements and incorporated classical principles into their design. These same principles of balance and proportion were used even though the style and appearance of the shopfronts differed.

Unaltered traditional style shopfronts that are typical and good examples of their time should be retained as should traditional features such as stall risers and cornices as they provide a valuable source of reference and identity in an area. Sensitive repairs are often preferable, but where necessary, features should be replaced as close to the original as possible. Where missing, features should be reinstated in a manner that is in sympathy with

20 世纪初期

一般来说，店面设计的既定原则没有变化，只是设计风格会随着店铺的类型进行调整。以肉铺和酒吧为例，它们的店面设计中易于清洗的釉面瓷砖就较为常见。

20 世纪末期

店面设计的核心理念在 20 世纪六、七十年代发生了彻底的改变，传统的店面设计不再流行。超大的橱窗，大面积的横带和廉价的建筑材料受到推崇，对原有建筑的无视使许多商业街失去了个性与韵味。

4.2 店面改造原则

在以往的店面设计中，位于一楼的店面通常被看做是建筑不可分割的一部分，因而一般在设计上都与上层建筑的结构和特点相互融合。

传统的店面设计通常采用的是相同的设计元素和一套经典的设计标准。即便店面的风格和外观不尽相同，设计师还是会一贯使用这些对平衡和比例起指导作用的设计标准。

未经改造的传统风格店面如果能够体现出一个时代的特色，则建议设计师对其进行保留，并作为当地一处宝贵的历史资源对其进行保护。设计师在设计过程中应选择谨慎小心的修复手段，

its surroundings. (See figure 4.4)

Classical Features

In traditionally designed shopfronts the classical details are sometimes incorrect or simply missing. Inadequate attention to traditional construction techniques and materials can ruin an otherwise attractive design.

A traditionally designed shopfront will have a timber architectural framework around the shopfront of pilasters, capital and plinth, console bracket, cornice, fascia and stall riser framing the display windows and giving visual support to the upper floors.

Various elements can be used to enclose the shop window and entrance. These include the fasçia, pilasters, cornice and stallriser, all of which has its own visual and practical function. A design based on the traditional shopfront incorporating these elements is always likely to be the most appropriate in an historical setting. However, it is the creative interpretation of traditions, which has led to a lively street scene, and the individual solution is encouraged.

Fascia

Surface mounted signs, which project from the historic fascias and partially conceal historic details, are not acceptable. They will not be permitted in conservation areas or on listed buildings. A traditional fascia comprising a painted timber background with hand painted lettering or raised timber letters is particularly appropriate on older buildings. Shiny or brightly coloured materials are rarely acceptable. Transfer lettering may be a suitable alternative to hand painted lettering in some instances. On traditional shopfronts it was

必要时应对这些特色元素进行最大程度的复原。如果特色元素已经缺失，则应该以一种与环境相符的方式将其重新呈现出来。（图4.4）

古典特色

传统店面设计案例中，一部分古典装饰细节是不恰当或者完全缺失的。如果设计师对传统建筑工艺和材料不够了解或重视，很可能毁掉一个原本十分成功的店面设计。

传统店面设计通常包含一个木质建筑框架，其融合了壁柱、护墙板、装饰性支架、挑檐、横带和竖板等组织结构。它一方面限定了橱窗的范围，另一方面在视觉上起到支撑上层结构的效果。

店面橱窗和入口处可以使用多种装饰元素，如横带、壁柱、挑檐和竖板。每个元素都有其独特的观赏和实用功能。在传统店面基础上结合了这些元素的设计方案往往是最适合店面历史底蕴的。然而，对传统进行创造性的解读，打造充满活力的新景观也是同样被鼓励的。

横带

在遗留下来的横带结构上直接安装突出店面的标识是不可取的。不应在保护区域或受保护建筑范围内进行这样的改造工程。涂漆木质背景上装饰手绘字母或突起标识的传统横带尤其适合位

common to use particular styles of lettering to reflect the actual use of the shop. The use of a 'house style' may be accepted if it does not detract from the character of the shopfront and its contribution to the street scene. As a general principle lettering should be painted directly onto the fascia in a colour that compliments that of the shopfront. Most traditional fasçias do not exceed 380mm (15") in depth.

Projecting box fascia signs are not appropriate on historic buildings and in conservation areas. They normally detract from the appearance of the shopfront and are over dominant in the street scene.

Cornice
This is traditionally timber in construction and provides a horizontal line between the shopfront and the upper floors and gives an element of weather protection to the shopfront.

Stall Riser
Historically, the presence and height of a stall riser varied with the nature of the goods on sale; for instance, some traders such as tailors and boot-makers required their goods to be viewed from above so that the stall riser was kept low. The context, scale and design of the shopfront will be an important consideration in the decision to incorporate or omit the stall riser in a new design. (See figure 4.5)

Access and Doorway
The traditional door is normally part glazed. The entrance door would be set back from

于历史建筑中的店面。有光泽的以及色彩明亮的装饰材料一般不建议使用。个性的店名设计必须与建筑和街道景观相匹配方能使用。一般来说，店名直接涂刷在横带上，配色与整个店面设计相呼应即可。多数传统店面的横带部分深度不会超过380毫米。

城市历史建筑和受保护区域内不适合使用突出的箱式横带。这种横带结构通常与店面外观属于不同风格，在街道景观中过于显眼。

挑檐
传统店面中的挑檐通常是木质结构，在水平方向上连接店面和上方楼层。同时挑檐还具有遮风挡雨的功能。

竖板
竖板的高度通常根据店内所售商品而有所变化，例如，裁缝店和鞋店通过橱窗展示的商品需要顾客从低处查看，竖板因而设计得比较低。设计师决定是否在一个新店面工程中使用竖板结构前，应对店面的环境、规模和设计风格等因素进行考虑。（图4.5）

出入口
传统设计中店面大门一般会使用部分玻璃。大门门口一般与人行道边缘保留一定距离，旧时遗留下来的台阶即便倾斜变形也

Figure 4.5 Olivocarne Restaurant, London, UK, design by Pierluigi Piu, photo by Pierluigi Piu Figure 4.6 Como quieres que te quiera, Buenos Aire, Argentina, 2012, design by The Wall Arquitectura y Diseño, photo by Julio Masri

图 4.5 橄榄树餐厅，英国，伦敦，皮耶路易吉·皮乌设计，皮耶路易吉·皮乌摄影
图 4.6 "如果我想要……"服装店，阿根廷，布宜诺斯艾利斯，The Wall 设计公司，胡里奥·马斯里摄影

the edge of the pavement. Historic steps should be kept even if ramped over. Recessed doorways are a common feature of traditional shopfronts. They allow for increased window area and larger display area. In the past many recessed doorways have been removed but the preferred option would be for those within Conservation Areas or in listed buildings to be reinstated. (See figure 4.6)

It is important in principle that disabled people should have a dignified easy access to and within historic buildings. If it is treated as part of an integrated review of access requirements for all visitors or users and a flexible and pragmatic approach is taken it should normally be possible to plan suitable access for disabled people without compromising a building's special interest. Alternative routes or re-organising the use of spaces may achieve the desired result without the need for damaging alterations.

Window

The shop window will typically include timber vertical mullions and a transom rail at door head height with transom lights above.

In some cases shop windows are filled with posters, stickers, temporary banners and illuminated signs. This type of advertising can detract from the overall appearance of the shopfront and can detract from the appearance of the street. On listed buildings this will normally require listed building consent and will generally be resisted.

Canopies and Blinds

If the building is a heritage asset or is within a conservation area, only fully retractable

应该尽量保留。凹进的门口在传统店面设计中十分常见，这种设计留出了更大的橱窗空间，具有更好的展示效果。近年的店面翻新工程中，许多凹进式门口被拆除改造，但是对于保护区域和历史建筑内的凹进门口，最好的处理方法是对其进行保留复原。（图 4.6）

在历史建筑内为残障人士增设通行设施是设计师应该重视的一个重要设计原则。将设计适合所有顾客的通行设施作为建筑景观规划的一部分，并且采用灵活实用的设计方法，既可以满足不影响建筑特殊功能和意义的条件，又可以实现方便残疾人通行的设计与改造。这一般会涉及到备选路径的设计和空间的重新规划。

橱窗

橱窗通常包括木质竖档和门头高度的横梁，横梁上方配有照明装置。

一些情况下，橱窗上会出现海报、贴纸、临时宣传横幅和灯光装饰。这种宣传手段会分散人们对店面整体以至街道景观的注意力。若在受保护建筑周边进行这种设计安排，通常需要提前获得许可。

遮阳篷和百叶窗

如果该建筑属于名胜古迹或者位于城市保护区内，则只允许使

Figure 4.7 Läderach Chocolatler Suisse, Baden Baden, Germany, designed by studio KMJ, photo by studio KMJ

图 4.7 Läderach 瑞士巧克力店，德国，巴登巴登，KMJ 设计工作室设计，KMJ 设计工作室摄影

canopies / blinds will be permitted. Dutch blinds and similar non-retractable blinds are inappropriate in conservation areas and on historic buildings. Dutch blinds and balloon canopies can look out of place too, whatever material they are made of and, will not normally be permitted. Traditional blinds made of canvas or similar non-reflective materials are to be preferred, especially on listed buildings and in conservation areas.

Pilasters and Consoles

Traditional shopfronts were often canted (or tilted) forwards and contained within the console brackets. Pilasters and consoles vary from being very elaborate and highly decorated to being relatively plain but they usually have some moulding or surface decoration. Where traditional pilasters and console details exist they should be retained. (See figure 4.7)

Entablatures

Entablatures were the forerunners of modern shop fascias. Typically they have a relatively shallow fascia topped by a generously moulded cornice, which provides a positive cap to the shopfront.

Colour Scheme

Dark, rich, primary matt colours are the most appropriate background colours for shopfronts on historic buildings and within conservation areas.

Materials

The materials used should complement the style and period of the building and the area. Generally, traditional shopfronts require the use of natural materials; styled, coloured, and finished in traditional ways. Timber sections for mullions and transoms are rarely square, colours are often recessive.

用完全可折叠的遮阳篷 / 百叶窗。不可折叠的百叶窗和遮阳篷对于该地区的店面来说会显得格格不入。建议选择传统的帆布百叶窗或由相似的不反光材料制成的百叶窗，在受保护建筑和区域内尤其如此。

壁柱和支架

传统的店面通常会朝向建筑正前方，店铺的两侧一般会设置装饰性的支架。壁柱和支架的造型相对比较多变，他们的风格既可以极度精美，也可以简洁流畅，但一般都具备一定程度的造型或表面装饰设计。对于已有的传统风格壁柱和支架元素应尽量保留。（图 4.7）

台口

台口是现代店面装饰中横带的前身，通常从较浅的横带上从容地延伸出挑檐，构成完整的店面造型。

配色

对于那些位于城市历史建筑和保护区域内的店面设计工程来说，比较适合选用较为深沉、浓重的亚光色彩作为店面的背景色。

建材

店面设计所选择的建材应与店面所在建筑的风格和时期，以及店面所在地区的特点相符合。一般情况下，传统店面比较适合选择天然环保建材，在设计上采用传统风格、传统配色和传统工艺进行修建。竖档和横梁所用的木材很少是方形的，色泽通常自然渐变。

Figure 4.8 Lindt Cafe Chapel St., Melbourne, Australia, designed by Rolf Ockert Design, photo by Rolf Ockert Design

图 4.8 查普尔街林特咖啡馆，澳大利亚，墨尔本，罗尔夫·欧克特设计公司设计，罗尔夫·欧克特设计公司摄影

Timber was the traditional shopfront material of previous centuries. It remains the most appropriate and versatile material. Timber is durable, versatile and inexpensive. It is also easy to maintain by painting. Timber along with other traditional materials such as brick and stone are the preferred choice of material especially for Listed Buildings and buildings in Conservation areas.

On Listed Buildings and quality buildings within Conservation Areas the use of non-traditional materials will not normally be permitted (i.e. fibre glass or plastics) as their appearance often conflicts with the character of the building and area.

Lighting

Where there is a concentration of listed buildings and where street lighting provides a subtle ambience, over-illumination can be garish and invasive and illumination may not be applicable at all.

Internally illuminated fascias and signs can create an unacceptable glare at night and are generally not acceptable. Individually illuminated box fascias, illuminated box signs, and individually lit Perspex letters will not normally be permitted.

Subtle and concealed lighting can be acceptable, depending upon location. The light source should be unobtrusive and should be directed carefully at the sign or shopfront to avoid glare and light leakage. (See figure 4.8)

Security

Large single sheets of glass should be avoided on traditional shopfronts. Laminated glass should be used for public safety, and as a security measure.

过去的几个世纪里，木材一直都是用于店面装潢的常见材料。直到今天木材也仍是最受人们青睐，并且便于设计的建材。平时仅涂刷油漆就可以对木材表面进行有效的保养和护理。受保护建筑和受保护区域内的建筑尤其适合使用木材与砖石等其他传统建材。

在受保护建筑中使用非传统建材（例如玻璃纤维和塑料）通常是不被允许的，因为这些非传统建材的外观往往与传统建筑和区域的特点相冲突。

照明

在受保护建筑集中的区域以及街道灯光较为柔和的区域，过度的店面照明可能会显得花哨而与周围环境格格不入。内部照明灯箱和广告牌的灯光可能在夜间显得过于刺眼，这种情况也不建议采纳。单独照明的灯箱、广告牌等通常也不允许在受保护建筑上使用。

可以采纳的是灯光柔和、半遮盖式的照明装置，款式和形式可以根据具体的位置来进行选择。光源应该尽量选择沉静柔和的类型，朝向店铺标识或店面，避免眩光并减少对其他建筑的影响。（图 4.8）

安保

传统店面应该避免使用单层大玻璃橱窗，而将夹层玻璃作为一种安保手段进行使用。

PART, 5

第 ❺ 篇　新店面设计法则

NEW SHOPFRONT
SIGN PRINCIPLES

Figure 5.1 ZARA Kumamoto, Kumamoto city, Japan, designed by Key Operation Inc. / Architects, photo by Itou Prophoto Corporation
Figure 5.2 Percimon, Medellín, Colombia, designed by Sofia Mora – María Velásquez, photo by Camila Mora

图 5.1 ZARA 熊本店，日本，熊木市，KEY OPERATION 股份有限公司／建筑师事务所设计，伊图摄影公司摄影
图 5.2 Percimon 酸奶店，哥伦比亚，麦德林市，索菲亚·莫拉，玛利亚·委拉斯凯兹设计，卡米拉·莫拉摄影

New shopfronts should be designed as an architecturally meaningful whole, not as an assemblage of different parts or separate elements. The size and shape of the shopfront should be determined by the size and form of the building it is to be located within. Some basic principles, such as the unity between shopfront and building, the control to scale, height and proportion, the choice of colour and materials, and the way of lighting should be taken into account when designing the new shopfront. A new replacement or altered shopfront should be sympathetic to the building to which it would be fitted and not detract from the character or appearance of the street. Advertising should be incorporated as an integral part of the design.

Also there are other elements having the impact on the new shopfront design, such as window positions, spacing and vertical or horizontal emphasis of the host building.

All new shopfronts and any alterations carried out to them should promote sustainability principles within their design, construction and performance. These principles should aim to improve energy efficiency and reduce impacts on climate change as well as reduce the use of materials and promote their re-use and recycling where appropriate. (See figure 5.1, 5.2, 5.3)

In brief, the keys for designing a new shopfront including:
• A new shopfront should be designed as part of the whole building;
• The design should reflect the context of the building and the street in which it is located;
• The design should take account of the vertical and horizontal emphasis of the surrounding buildings and the street scene;
• The Scale, height and proportion of the shopfront should take account of the basic

设计师在对店面进行设计时，应当把新店面看做一个有意义的建筑结构单元，而非不同独立元素的组合。设计师只有在充分了解店面所处大环境的基础上，才能对其进行设计改造。店面的大小和形状应当由它所在建筑的规模和形式决定。店面设计的基本原则包括，店面与所在建筑的整体统一，对店面大小、比例和高度的把握，店面的配色和建材的选择，以及对照明装置的设计。翻新或者改造店面的工程应当与所在建筑的风格相符合，避免与街景形成不适宜的反差。宣传手段也应作为设计的一部分融合在店面之中。

橱窗位置的设定、间隔和建筑在竖直及水平方向的视觉重点等其他元素也对新店面设计有一定的影响。

与新店面设计和改造工程相关的设计、施工和使用都应遵循环保的原则，努力提高能效，减少气候变化带来的影响，减少材料使用，促进物资的回收和再利用。
（图 5.1，图 5.2，图 5.3）

简而言之，新店面设计中应该注意以下几点：
· 将新店面作为建筑的一部分进行设计
· 设计方案应该反映所在建筑和街道的特点
· 设计方案应考虑到周围建筑和街景在竖直和水平方向上的重点视觉元素
· 在设定店面的大小、比例和高度时应当充分考虑到基本的建

Figure 5.3 NICKIE in Lishui, Lishui, China, designed by SAKO Architects, photo by Ruijing Photo
图 5.3 丽水尼基儿童服装店，中国，丽水，迫庆一郎事务所设计，瑞金摄影

architectural elements.
- Detailing should be appropriate to the age and character of the building;
- The design should ensure access for all;
- The design and construction should take account of sustainability;
- Security should be an integral part of the design of shopfront.

5.1 Unity between Shopfront and Building

Alterations and changes to shopfronts occur over time and are often carried out to suit different traders and occupiers. The overall effect of these changes, can over time, have an adverse impact on the shopfronts concerned and cumulatively when several premises are involved impact on the character of town and district centre shopping areas.

It is important that changes carried out to shopfronts, whether alterations or replacements are guided to ensure sensitive and sympathetic schemes which enhance the character of not only individual buildings but the wider street scene and make positive contributions to trading and social success.

In more recent years the mass produced 'one style fits all' shopfront has widely used. This type of mass produced design takes no account of the building in which it is located. Today the use of standard shopfronts and inflexible corporate styles is in danger of eroding the character of towns and district centres.

Whether designing a new shop or planning to alter or replace an existing one, it is important

筑元素
- 店面的细节设计应该符合建筑的年代和特点
- 设计方案应该考虑到特殊人群的通行需求
- 设计和施工应该考虑到环保的问题
- 店面设计应该将安保作为重要的考虑因素

5.1 店面与所在建筑的统一

随着时间的推移，经常需要对店面进行调整和改造，以适应店主以及经营范围的更替。改造后的局部结构可能逐渐对店面整体产生负面的影响，而当多个店面进行了同类改造时，影响可能涉及整个商业区。

对已有店面进行改造的目的就是改造工程能够使单个建筑以及整个街景的特点得以强化，起到促进经营和社会发展的积极作用。在设计时需要采用细致周到的方法实现改造是设计师应注意的重要问题。

近年来流行"批量式"的店面设计。这种设计模式欠缺对店面环境因素的协调性和针对性。目前人们已经逐渐意识到这种缺乏灵活性和适应性的店面设计模式对城市景观和商业区景观有着一定的消极影响。

无论是设计新店面还是对已有店面进行改造，设计师应考虑预

Figure 5.4 Levi Strauss, Berlin, Germany, designed by Checkland Kindleysides, photo by Checkland Kindleysides
Figure 5.5 Casa Turia, Valencia, Spain, 2013, designed by CuldeSac™, photo by CuldeSac™
图 5.4 柏林里维斯旗舰店，德国，柏林，切克兰德·金德利赛斯设计咨询公司设计，切克兰德·金德利赛斯设计咨询公司摄影
图 5.5 Casa Turia 红酒店，西班牙，瓦伦西亚，CuldeSac 设计公司设计，CuldeSac 设计公司摄影

to consider the visual impact that will have upon the building and the wider streetscape. A shopfront and building must be considered as a whole entity to ensure they are seen together rather than separate elements of the same building.

Good design and a high quality environment go hand in hand. A carefully designed and eye-catching shopfront is good for business and can make a positive contribution to the character of the street and the vitality of the retail areas. Conversely, a poorly designed shopfront can be visually intrusive and harm the retail area. An attractive shopping street is good for all.

In the past, most shopfronts were designed as an integral part of the building and based on classical proportions, with the various elements forming a balanced composition with the building. While today, there are some excellent shopfronts in city, others have been harmed by unsympathetic alterations and are out of keeping with the building and the street scene. This chapter does not set out to prescribe specific styles and is not intended to restrict ideas, but to encourage appropriate high quality sympathetic design. The style which a new shopfront should take will vary depending on the age and type of the building. A well designed shopfront will complement the building and enhance the character of the street. (See figure 5.4, 5.5, 5.6, 5.7)

Each street has a character and visual hierarchy that is established by the relationship between the buildings found there. If a shopfront is to be successfully integrated into its surroundings, it will be required to respond to a number of established design criteria. Generally, new and refurbished shopfronts should follow the principles set out below:

期效果在视觉上对所处建筑和街区景观的影响。设计师应将店面与所在建筑作为统一整体进行考量，以便作为一个完整的景观而非独立的元素呈现出来。

良好的环境与合理的设计密不可分。精心设计出的店面引人注意，利于业务的发展，对所在街道景观的塑成以及保持商业区的活力有着积极的推动作用。同样，不科学的店面设计会引起消极的视觉效果，不利于商业区的发展。创造出有吸引力的购物商业街才是所有人都期待和向往的。

过去的设计实践中，绝大多数设计师会将店面设计作为建筑的一部分处理，依据经典的设计比例，利用多种不同建筑元素构成与整个建筑和谐的店面设计。但如今，尽管商业区分布了不少出色的店面作品，还是会有一些店面由于不适宜的改造工程而显得与周围环境格格不入。本章内容并非围绕特定的店面设计风格展开，也不希望制约到设计师的想法，而是为了鼓励设计师从事高品质的和谐设计。各个新店面的设计风格会依据所在建筑的年代和类型而有所区别。精心设计的店面会与建筑形成互补，对街道景观起到加强作用。（图 5.4，图 5.5，图 5.6，图 5.7）

每条街道的视觉层次和特点都是在周边建筑网络的基础上构成的。想让一个店面成功地融入周围的环境，通常需要满足几个条件。以下的几项设计原则适用于大部分的新店面设计工程以及店面翻新设计工程：

Figure 5.6 ARMAZEN, São Paulo, Brazil, designed by ODVO Arquitetura e Urbanismo, photo by Pregnolato e Kusuki estúdio Fotográfico
Figure 5.7 Permy-mi Jang Won, Gyeonggi-do, South Korea, designed by M4, photo by M4
图 5.6 ARMAZEN 食品店，巴西，圣保罗，ODVO 建筑事务所设计，普雷尼奥拉托与玖月摄影工作室摄影
图 5.7 Permy-mi Jang Won 美发沙龙，韩国，京畿道，M4 设计公司设计，M4 设计公司摄影

About the street scene

The main point to consider in the design or alteration of a shopfront is how the building fits into the street. The width of the buildings and their height make the character of the street. There might be a vertical or horizontal emphasis to the architectural features. This is the rhythm of the street, and where a shopfront extends across several different buildings, the rhythm of the street can be spoiled. If the buildings differ in size or architecture varied shopfront designs are likely to be more appropriate.

- Reinforce the local character of the area and contribute to or create a sense of place;
- Shopfront design towards enhancing the overall quality of the street scene;
- Consider the impact of the design on the character of the street;
- Proportions, materials and details should maintain and reflect the variation of nearby buildings. The shopfront should not dominate its surroundings.

About the building

A shopfront should relate to the building it belongs to so that it forms an integral part of the elevation rather than an isolated element on the ground floor. This can be achieved by taking account of the scale and architectural style of the buildings and by echoing the arrangement of windows and areas of walling on the upper floors.

- Consider the shopfront as part of the whole building;
- A well designed shopfront will harmonise with the style and proportions of the building. Good guidance can be obtained from looking at the style and proportions of the building and any surviving fabric and historic photographs, looking at neighbouring buildings and

关于街道景观

店面设计和改造过程中需要考虑的一个主要问题是改建后的建筑如何融入街道景观。建筑的宽度和高度构成了街道景观。在众多的建筑元素之中，会存在一个竖直或水平方向上的视觉重心。这也是街道景观的韵律。如果店面横跨多个不同类型的建筑，则会打破这种韵律之美。如果涉及的建筑属于不同大小和风格，那么分别采用相应的店面设计方案会更为适合。

- 强化当地特色，构成或提升空间感
- 加强街道景观的整体质量
- 考虑到设计方案对街道风格和特点的影响
- 设计比例、建材和细节应保留并尊重附近建筑的多样性，而不应成为周边环境的支配因素

关于店面所在建筑

店面应与其所在建筑建立联系，从而构成一个统一的立面有机体，而不是独立的一个楼层。在对其进行设计时应当参考建筑的大小和风格，在橱窗和墙体设计上进行呼应，有助于实现这一目标。

- 将店面作为整个建筑的一部分进行设计
- 经过精心设计的店面不仅可以提升自身品牌形象，还可以协调整个建筑的风格和比例。参考建筑的风格和比例，现存的织物，老照片，相邻建筑和同一区域其他协调的店面设计案例是比较

PART 5 New Shop Front Design Principles

other sympathetic shopfronts in the area;
• Conserve existing historic features where possible, this may be essential if the building is listed;
• Use symmetry and rhythm to relate shopfronts to upper floors, unless a strong horizontal emphasis exists to allow a different pattern;
• Adapt company 'house styles' to suit the individual character of the area and building, avoid using arbitrary repetition and stretched logos.

5.2 Control to Scale, Height and Proportions

The scale, height and proportions of a shopfront should be in proportion with the building. The shopfront and any upper floors should work together, rather than separately. This will allow the design of the shopfront to fit within the original structural framework of the building.

In small-scale buildings, the shopfront should also be small. The size of the display windows, the depth and height of the fascia and the proportions of the details should all be modest. In larger buildings the shopfront can be larger but still in proportion to the building. Over-large fascias are the most common disfiguring element of existing shopfronts and they often obscure important architectural details. Where excessively deep fascias have been introduced in the past, the overall height should be decreased in any replacement.

Where a shopfront is to cover more than one building, vertical subdivisions should be used to retain the individual appearance of each. This can be done by retaining dividing piers

有效的方法
・具有历史价值的元素应尽量予以保留，对于受保护建筑更是如此
・通过对称和重复的手段将店面与建筑其他部分联系在一起，原建筑具备明显的水平视觉重心，适合采用其他设计模式时应另当别论
・特定品牌的标准化店面设计需要因地制宜，以适应所在建筑和区域的特点。应避免刻意重复和过于夸张的品牌标识

5.2 对店面大小、比例和高度的把握

设计时对店面的改造并不是凭空想象出来的，店面的大小、比例和高度都应该参考所在建筑的相关数值。整个店面应当与上层建筑融为一体，这就需要设计师将店面设定在整个建筑的框架内。

小型建筑中的店面也应该设计得相对小一些。橱窗的尺寸，横带的深度和高度，装饰细节的比例都应控制在一定范围内。大型建筑中的店面应相对较大，但仍需与建筑构成适宜的比例，过大的横带是已有店面中最常见的不和谐元素，往往会遮挡住重要的建筑细节。针对已有店面中过深的横带，建议设计师在改造工程中将横带高度减少。

当店面横跨多个建筑时，设计师应注意在竖直方向上增加区域

Figure 5.8 Puma Store London, London, UK, designed by Plajer & Franz Studio, photo by Mmanuel Schlüter/ copyright Puma AG

图 5.8 伦敦彪马旗舰店 , 英国 , 伦敦 ,Plajer&Franz 工作室 , 曼纽尔•施吕特尔摄影 / 版权归彪马品牌所有

or pilasters, and respecting differences in adjacent fascias and stall risers. Individual fascias should be retained; conformity and linkage can still be achieved satisfactorily by continuity of lettering style and design. (See figure 5.8)

Where a shopfront covers more than one building or façade, shopfronts should be individually designed for each unit of façade. Where a shop straddles two different buildings the shopfronts can abut on the line of the party wall with a double pilaster and use common colour schemes and materials.

Poor quality shopfronts can erode local character and provide an unattractive place for visitors. Long unbroken shopfronts do not respect the character of the building and have little visual appeal. Well designed shopfronts improve the shopping experience and enhance their surroundings. A shopfront can be enhanced with a sympathetic design and restore the architectural unity of the building.

Many buildings in shopping areas are symmetrical and to keep a sense of balance, this symmetry should be extended to the shopfront. Sometimes internal planning makes it difficult to achieve exact symmetry, but often a compromise is possible to enable a satisfactory outcome. Intermediate columns and window mullions can contribute some visual balance.

Many C20 shopfronts have large expanses of glass, which make the building above it appear unsupported. This can look particularly uncomfortable if the shop window straddles two or more buildings. Columns, pilasters or areas of walling can be used to give the building visual strength.

划分的设计，以便保留各个建筑的特色。具体的方法可以采用独立窗间壁或壁柱，呼应邻近的横带和竖档。独立的横带应尽量保留。店面的一致性和连续性可以通过统一的字体和设计来实现。(图 5.8)

店面横跨多个建筑或外墙的时候，设计师应对不同建筑的店面分别进行设计。如果店面跨越两个不同类型建筑，应将店面设计成紧靠界墙，采用双壁柱形式，并且使用常见的配色和建材。

粗糙的店面设计会削弱该区域原有的特色，对街道景观的提升起到负面作用。过长而没有间隔的店面设计会使所在建筑失去个性，视觉上也不甚美观。成功的店面设计则能够改善整体购物环境，提升消费者的购物体验。在和谐的店面设计中，店面本身的个性也会得到强化。

商业区的许多建筑都是对称、平衡的。这种对称感也要在店面设计中加以体现。有时在规划方案中实施绝对对称有很大的难度，但只要做出一点点妥协，就可以获得满意的结果。合理使用长度适中的柱子和窗间小柱对获得视觉平衡有一定的效果。

许多 20 世纪店面有大块的玻璃橱窗，这种设计会让上方建筑显得摇摇欲坠。如果店面横跨多个建筑，则更会造成视觉的压迫感。多种形式的柱子和墙体可以改善这种状况，给建筑带来视觉上的支撑。

Figure 5.9 Camper store, Granada, Spain, designed by A-cero Joaquin Torres & Rafael Llamazares architects, photo by Juan Sánchez
Figure 5.10 Farm Direct, Hong Kong, China, designed by Wesley Liu, photo by Wesley Liu
图 5.9 格拉纳达 Camper 鞋店，西班牙，格拉纳达，A-cero 杰奎因•托雷斯和拉斐尔•利亚马萨雷斯建筑师事务所设计，胡安•桑切斯摄影
图 5.10 "农场直达"水培菜零售店，中国，香港，廖奕权设计，廖奕权摄影

5.3 Colour Scheme

Colour is a very important consideration. When considering the colour of new or replacement shop-fronts it is important that the colour scheme complements the character and appearance of the building rather than conflicting with it. Standard corporate colour schemes should be adapted to suit the character of the area. Any proposed colour scheme should be in keeping with the existing colour scheme on the building and adjoining buildings. It should enhance the design of the shopfront and highlight any important decorative features present. The range of colours used should generally be kept to a minimum.

The sensitive use of colour offers much scope for improving the street scene. Harsh or gaudy colours draw undue attention to themselves and should be avoided. Rich dark colours look very good as they leave the window display to provide the highlight. Off-white is also a traditional colour. (See figure 5.9, 5.10, 5.11)

The imposition of a corporate colour scheme regardless of the location may erode the character of an area, but minor variations of the corporate colour e.g. just a small proportion of the fascia in house colours might not alter the ambience of the street. A single colour should be used for all major elements, with a contrasting colour picking out key features to good effect. Lighter colours such as white and cream as the main colour should be avoided as they discolour easily. Large areas of bright colour, particularly on shiny materials such as plastic should also be avoided. The use of garish or vivid colours, not in keeping with the character of the area will not be appropriate and will not be supported.

5.3 配色

色彩搭配也是店面设计中需要设计师考虑的一个重要因素。设计师在接收一个新项目设计时，无论是对新店面的设计还是对旧店面的改造，都应当注意让配色方案与建筑的特点和外观相呼应，以避免出现色调不和谐的设计。配色方案还应当与所在建筑及周围建筑的配色相符合，同时突出店面的设计，强调已有的重要装饰元素。所使用颜色的种类应尽可能少。（图 5.9，图 5.10，图 5.11）

合理的色彩搭配不仅可以给人们带来舒适的视觉感官，而且还能营造出令人愉悦的街道景观。过于明亮和华丽的颜色会吸引不必要的注意力，设计过程中应予以避免；而浓厚的深色则可以彰显品牌的高贵华丽，还为橱窗的展示留出了更多的空间，建议使用。另外，纯白色也是常用的传统店面配色。

店面配色必须要考虑店面所处的地理位置，如果不能做到因地制宜，而是简单照搬企业色彩方案，通常会对这一区域的风格形成干扰。在企业色彩方案的基础上做微小的调整是被广泛提倡的，例如占横带较小比例的某种颜色与建筑原本的某种配色保持一致，这样可以尽量减小对街道景观整体效果的影响。建议选用同一种颜色作为店面的主色调，而重要的装饰元素则可以使用一种对比色来进行突出加强。设计师应避免选用白色和米色等容易褪色的色彩作为店面的主色调。在对店面进行设计

Figure 5.11 Mango Financial, Texas. USA, designed by Bercy Chen Studio LP, photo by Ryan Michael
图5.11 芒果金融，美国，得克萨斯州，贝尔西·陈工作室设计，瑞安·迈克尔摄影

Elements such as mouldings to doors and pilasters, console brackets and fascia details are examples where lighter shades/hues have been used. Picking-out colours are chosen to harmonise and compliment the principal shopfront colour rather than to make a stark contrast to it. A picking-out colour scheme is not being prescribed, as in the same form of suggested base colours, because of the scope to use complimentary hues, white and even metallic gold paint. Windows, not forming a shopfront and those above ground floor level, should not be painted to compete with either the ground floor colour scheme or the wall within which the particular window is set. The window should present a softer, complimentary colour taking its cue from the brick or render wall colour, not set as a contrast. Timber shopfronts should be painted and not stained or varnished.

5.4 Lighting

As a general rule, most illuminated signs will require local Advertisement Consent.

Lighting a shopfront is normally positive, as it can help create a safe and visually interesting night time environment. However, street lighting and lighting from window displays are often quite sufficient for providing attention.

It is possible to successfully incorporate lighting into a shopfront design. Deciding on the most appropriate method, design and type of lighting will be the key to ensuring it is not detrimental to the character or appearance of the building or area.

Modest and subtle lighting can add sparkle and vitality to the night-time scene. In all of the

时，还应该注意避免大面积使用明亮的色彩，底部材料是塑料等反光材料时尤其如此。那些与所在地区风格不符的花哨、鲜艳的色彩也应谨慎使用。

行业中对店面使用的色彩范围没有固定的要求，但是也不能随意的使用色彩。门口和壁柱旁的装饰，支架和横带的细节等都是适合较浅颜色的部位。其他部分的色彩应与店面主色调协调、互补，避免强烈的反差。橱窗应该选用柔和的互补色，与砖石或墙壁颜色相近，尽量避免对比色的使用。木质店面应该用有色油漆处理，避免污迹和清漆。

5.4 照明

总的来说，大部分灯光装饰在使用前需要首先获得当地相关部门的批准。

店面照明一般起到积极的作用，因为它有助于营造一个安全、个性的夜间环境。但是街道照明和橱窗照明通常已经足够明亮，可以充分地吸引行人的注意力。

将照明装置成功地添加到已有店面中是可行的。设计师应选择最适宜的照明装置设计方法和照明类型，以确保适应店面所在的建筑及地区。

Figure 5.12 International Giolitti, Istanbul, Turkey, designed by NABITO ARCHITECTS, photo by NABITO ARCHITECTS
图 5.12 焦利蒂冰激凌店，土耳其，伊斯坦布尔，NABITO 建筑师事务所设计，NABITO 建筑师事务所摄影

market towns though, restraint is required.

Illumination of the fascia needs to be given careful thought and be sensitively incorporated into the shopfront composition. There are two basic ways of illuminating fascias; either internally through box signs or externally by means of spotlighting or strip-lighting.

The preferred choice of lighting is external lighting of the fascia. Where external lighting is proposed and is appropriate for the building, it should be subdued, discreet and sympathetic to the building and the surroundings. The use of intermittent light sources, moving features, exposed cathode tubing or reflective materials are not considered acceptable lighting solutions. External lighting may be appropriate by the use of trough lights with a hood finish to match the background colour of the fascia.

This should be by means of concealed lighting such as slimline LED trough lighting (preferably recessed into a projecting cornice). Carefully positioned small spotlights may be an alternative. Large spotlights, swan neck lamps or heavy canopy lights should be avoided, as they can clutter a building and be over-bright. In all situations, only the lettering to a sign, and not the whole fascia, should be illuminated.

Full Internally illuminated fascia box signs and projecting signs are not in character with most retail areas in the district and will not be an acceptable form of illumination as this is almost always visually dominant. Individually illuminated box fasçias, illuminated box signs, and individually lit Perspex letters will not normally be permitted. More subtle forms of lighting include backlit lettering, individual halo letters and cold cathode tubes where only the

柔和适度的照明能够为夜景增添光彩和活力。但在大部分城市，市政部门对照明有一定的限制。

横带照明设计是需要投入较多精力进行设计的一个部分，目的就是能够使其较好地融入店面结构。横带照明有两种基本方式，灯箱内部照明或在外部使用射灯和条形照明灯。

横带外部照明是设计师们首选的照明方式。但是在使用中应当注意，外部照明应当温和、低调，并且能够与建筑及周边环境相称。不建议使用间歇性光源、移动灯光装饰物、外露的阴极管照明和反光材料。适当使用灯槽和灯罩可以使外部照明装置更好地与横带背景色相呼应。

LED 照明槽等隐藏式照明装置（嵌入到突出檐口为最佳）可以被用作店面照明。这里也可以选择位置适宜的小射灯加以替代。在进行店面设计时，要避免使用大射灯、鹅颈灯和重型雨棚灯，因为这些照明装置亮度过高，而且会使建筑外观显得杂乱无序。一般情况下，照明对象只包括标识的文字部分，而无需照亮整个横带结构。

由于外观过于庞大，与大部分商业区的风格不符，不建议在店面设计中使用内部照明的横带灯箱。单独照明的灯箱、广告牌等也通常不允许在店面中使用。较为不照亮背景、只照亮标识文字的柔和照明方式包括背光式标识、独立光环式标识和冷阴

Figure 5.13 ODESSA Restaurant, Kiev, Ukraine, designed by YOD Design Lab, photo by Andrey Avdeenko
Figure 5.14 Schutz Oscar Freire, São Paulo, Brazil, designed by be.bo., photo by Marcos Bravo

图 5.13 奥德萨餐厅，乌克兰，基辅，YOD 设计工作室设计，安德烈·阿夫迪恩科摄影
图 5.14 奥斯卡弗莱雷街舒兹店，巴西，圣保罗，be.bo. 设计公司设计，马科斯·布拉沃摄影

lettering and not the background is illuminated.

It may be more appropriate in some instances to consider the individual illumination of letters through halo background lighting. If this method is proposed the sign fascia should not be internally illuminated. For this method to be supported high quality design and materials will be necessary. Individual internally lit letters and halo lit letters outside a conservation area can be a subtle form of lighting, providing the letters have a small projection off the fascia.

Signage lighting, where it is not detrimental to the building and surroundings, will be acceptable where this takes the form of discreet / recessed LED trough lights in a cornice or a small number of spotlights or internally illuminated lettering or halo lighting behind individual letters and where the letters have a small projection.

On hanging signs, if illumination is appropriate for the building or area this should be through discreet slimline LED lights attached a short distance, such as 80mm, off the bracket arm.

Lit window displays can have a positive impact on the quality of the retail area and create a sense of security for users. Carefully illuminated windows displays using discreet light fittings can be attractive outside trading hours. Where a shop is lit overnight for security, the shop window should be illuminated from inside, and not the fascia. (See figure 5.12, 5.13, 5.14)

5.5 Choice of Materials

The shopfront design should propose high quality materials and finish. High quality, durable

极电子管照明。

一些情况下，在设计中选用独立光环式标识可能更为适宜。为了与之配合，横带标识则不宜采用内部照明的方式。精巧的设计和高质量的建材会使这种照明装置更加完整协调。非保护区域内可以选择独立的内部照明字母与独立光环式照明字母，作为一种柔和的光源使用。

与挑檐结合的内嵌式照明或少数几个射灯或内部照明字母或独立光环式字母等照明装置都可以在店面中使用，前提是对所在建筑和区域不会构成任何负面影响。

配合悬挂式标识的照明装置适宜选用隐藏式 LED 灯，设置在距离托架较短距离内，80mm 左右。

有照明设计的橱窗对商业区的视觉形象有提升的作用，同时明亮的照明也能够加强人们的安全感。良好的照明设计会使橱窗在营业时间以外仍旧散发魅力。如果出于安保考虑，橱窗需要彻夜照明，则应选择内部照明，而非横带照明。（图 5.12，图 5.13，图 5.14）

5.5 建材的选择

店面设计注重高质量的建材和表面处理。高质量、耐磨耐用的

and selected materials will be required in any shopfront design. The finish should enhance the shopfront design. These should harmonise with and complement the building. (See figure 5.15, 5.16)

The number of different materials and colours should be kept to a minimum to avoid a clash with the adjoining buildings and the overall character of the street.

Painted timber should be the basis of new designs in conservation areas and listed buildings. Timber is one of the most adaptable materials used in shopfront construction. It can be worked to almost any profile, is durable and can be freshened up by repainting at a minimum cost. Also, it is easy to maintain by painting. New timber, especially hardwoods, should be from legal and sustainable sources.

Generally speaking, glossy surfaces, acrylic or Perspex sheeting, aluminium or plastic shopfronts are not acceptable in historic areas. However, modern materials can be accepted where they are used with care and it can be shown that they will enhance an area. Where it is demonstrated that they preserve or enhance the character or appearance of the area and are not detrimental to the character of the building on which they are proposed, they can be considered. Other materials may be considered appropriate in specific locations or designs within the district. For example, metal frames, marble and brick may be appropriate if the building design and age lean towards the use of this type of material.

The use of UPVC material should be avoided as this will normally detract from the architectural quality of the building and character of an area. This material is normally not as cost effective as timber or aluminium shopfronts.

精选建材是每个店面工程必需的组成部分，表面处理则应该起到强化店面设计的效果。这两个方面都应与店面所在建筑构成协调补充的作用。（图5.15，图5.16）

所用材料与色彩的数量应控制在最低，避免与相邻建筑或整个街道景观形成冲突。

在受保护区域和建筑内开展新店面设计，设计师应将涂漆的木材作为主要建材。木材是店面建筑工程中适应性最强的一种建材，不仅可以改造成任何形状，具备持久耐用的特质，可以通过重新涂刷进行廉价改造，还易于保养维护。选择新木材，特别是硬木时，应该确认木材从合法环保的渠道获得。

总的来说，在受保护建筑中进行的店面设计要避免使用光滑的面板，丙烯酸或有机玻璃板材，铝或塑料材料。但是如果能够合理使用，为周围环境增添特色，现代材料的使用也是可以接受的。特定地点或特别的设计方案中可以适当选用其他材料。例如，如果建筑本身的设计风格或年代适合金属框架、大理石和砖石元素，则可以采用这种元素。

店面设计中应该避免使用未增塑聚氯乙烯材料，因为这种材料通常会降低所在建筑的质感，削弱该区域的特点。相比而言，在店面设计中使用木材或铝材的性价比更高。

Figure 5.15 Family Center, The Store, Noor,Iran, designed by Ali Alavi, photo by Ali Alavi

Figure 5.16 Deborah Milan Flagship Store, Milan, Italy, designed by Hangar Design Group, photo by Hangar Design Group

图5.15 家庭中心服装店，伊朗，努尔郡，阿里·阿拉维设计，阿里·阿拉维摄影

图5.16 黛博拉米兰旗舰店，意大利，米兰，Hangar设计集团设计，Hangar设计集团摄影

PART 5 New Shop Front Design Principles

PART, 6

DES
ARCHITEC

第 ❻ 篇　店面建筑元素的设计方法

Figure 6.1 DC Store Shinsaibashi, Osaka, Japan designed by Japan Specialnormal Inc., photo by Koichi Torimura
Figure 6.2 Quiksilver Lifestyle Store, Peru, designed by Marcia Arteta Aspinwall & Fausto Castañeda Castagnino, photo by Zander Aspinwall
图6.1 心斋桥DC店，日本，大阪，Specialnormal股份有限公司设计，鸟村弘一摄影
图6.2 极速骑板服装店，秘鲁，马西娅·阿尔特塔·阿斯平沃尔，福斯托·卡斯塔涅塔设计，桑德尔·阿斯平沃尔摄影

The design of a shopfront should be based on a number of key architectural elements which link together to form a visual and functional framework. The main elements include: pilasters (for visual separation between shopfronts), a cornice (for visual support and enclosure) and a stall riser (for a visual base); and a fascia. Other influential elements include: windows, access, canopies or blinds, and the displays on highways or footpath.

6.1 Fascia and Signage Design

6.1.1 Fascia and Detailing

The fascia is usually the most prominent feature on a shopfront as it provides the space for advertising and so will be designed to attract attention. It should usually be separated from adjacent fascias by pilasters, or some other form of vertical division. It should not extend, uninterrupted across a number of buildings, even if they are in the same ownership. Nor should they obscure other architectural details such as cornices, or upper storey windows. (See figure 6.1, 6.2)

The scale of the fascia should be appropriate to the character, height and period of the building and in proportion with the design of the shopfront. It should be well proportioned, and typically be no deeper than 1/5th of the height of the shopfront. The top of a fascia should be sited well below the sill of the first floor windows. A fascia should not obscure any existing architectural feature or decoration, extend above the ground floor ceiling level or across more than one building thant has a positive impact on the street scene. (See figure 6.3)

店面设计以几个主要的建筑元素为基础，相互关联构成一个视觉与结构的整体。这些主要设计元素包括：壁柱（在店面中起到视觉分隔的作用），挑檐（视觉上起到支撑和包围的作用），竖板（作为视觉上坚实的基础），以及横带。其他重要的元素还包括橱窗、门、遮阳篷和遮盖，以及放置在公路和人行道上的宣传展示。

6.1 店面横带和标识设计

6.1.1 横带及装饰细节

店面的横带适合放置广告，通常采用引人注意的设计，是店面设计中最为突出的一个部分。不同横带之间应当由壁柱或者其他形式的竖直分割元素进行界定。属于同一个店铺的横带不应直接跨越多个建筑，同时也要避开挑檐、楼上的窗口等其他建筑元素。（图6.1，图6.2）

在进行店面设计时，应当注意横带的尺寸大小应当与所在建筑的风格特征、建筑高度和建造年代等元素相对应。比例适中的横带设计通常不会超过整个店面高度的五分之一。横带的顶部应当在主建筑二楼窗口之下，并与之保持一定距离。横带设计应避开原有的建筑和装饰元素，在一楼天花板的高度横向延伸。（图6.3）

Figure 6.3 ASK Italian Hertford, Hertford, UK, designed by Gundry & Ducker Architecture, photo by Hufton + Crow
图 6.3 德特福德 ASK 意大利餐厅，英国，德特福德郡，甘德利 & 达克建筑公司设计，霍夫顿与克劳摄影公司摄影

However, a common fault is to make the fascia too deep in an attempt to maximise signage space. Oversized or deep fascias can have a heavy clumsy appearance. They harm the proportions of a shopfront and are often used to conceal suspended ceilings within the shop. Oversized fascias being used to conceal a suspended ceiling inside the shop will not be appropriate. If a deep fascia has been installed in the past, an opportunity should always be taken to improve the situation. Oversize and garish fasçias can be one of the most unattractive features of shopping streets. This can spoil the appearance of buildings by obscuring stringcourses and first floor windowsills as well as hiding features of original shopfronts.

Other methods must be considered to hide a false ceiling. For example, setting the suspended ceiling back from the window or forming a splayed bulkhead. Where a false ceiling is proposed inside a shop, it will not be acceptable to increase the depth of the fascia in line with this. The change in level can be overcome through transom lights with opaque glass or setting the suspended ceiling back inside the shop and splayed.

A fascia should not extend beyond the shopfront surround and should not stretch uninterrupted across a number of distinct buildings or architectural units, even where they are occupied by the same business. Conversely, where two users occupy the ground floor of a single building, the shopfronts and fasçias should be broadly co-ordinated.

Modern boxed fascias which project forward of the face of the building are often bulky and detract from the appearance of the shopfront. They have become heavily standardised and use aluminium frames and bright acrylic panels. And they are usually unsympathetic to the street scene. They are not permitted in conservation areas or on heritage buildings. If one is

横带设计中比较容易出现的错误是为了盲目扩大店面的广告空间而将横带设计得过大，这往往会对店面比例起到不好的作用。用超大的横带结构遮挡住店内吊顶也是不适合的做法。如果是店面原有的横带结构出现了这类问题，设计师应及时发现问题，并对其进行积极的改造调整。超大的花哨横带会遮挡束带层和二楼的窗沿，掩盖建筑原有的特色元素，对购物区景观构成最直接的影响。

如果出现特殊情况，有遮挡吊顶的需要，设计师则可以考虑使用其他设计手段。例如，让吊顶远离窗口，或者添加八字形隔板等。店内需要安装吊顶时，横带尺寸无需随之改变。使用不透明玻璃气窗或将吊顶位置内调可以克服这种结构上的高度变化。

同一个横带结构，即便属于同一个店铺，在设计时也不应直接跨越多个风格不尽相同的建筑单元。相反，如果是同一个建筑内的一层被两个不同的商家共用，店面和横带设计也应该尽可能协调。

现代化的箱式横带在现如今越来越多的被设计时所采用，这种箱式横带依附于建筑外墙，并向外突出，看起来非常的庞大笨重。如今这类横带的使用已经非常的规范化，铝框架和亚克力板材是最为常见的材质。由于它们色彩非常的鲜艳丰富，与街道景观通常不太协调，所以这种类型的横带不能在受保护区域和建

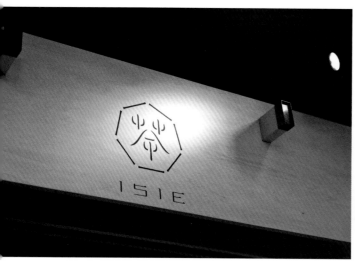

Figure 6.4 151E, Fukuoka, Japan, designed by PLANNING ES, photo by Hiroshi Mizusaki
Figure 6.5 El té – Casa de Dhás, Rio Grande do Sul, Brazil, designed by Gustavo Sbardelotto/estudio 30 51, Mariana Bogarin, photo by Marcelo Donadussi
Figure 6.6 Ray-Ban Store, Buenos Aires, Argentina, designed by Arquitectura y Diseño, photo by Julio Masri, Alejandro Hazan

图6.4 151茶铺，日本，福冈，ES设计公司设计，水崎浩摄影
图6.5 茶之家，巴西，南里奥格兰德，古斯塔夫·斯巴德洛托和马里亚纳·伯加林设计，马塞洛·唐纳杜斯摄影
图6.6 雷朋眼镜店，阿根廷，布宜诺斯艾利斯，架构与设计工作室，胡里奥·马斯里，亚历杭德罗·哈赞摄影

used on a modern building, it should be recessed behind the fascia.

Any detailing used to frame the fascia; for example a cornice, blind box, and/or corbels should be incorporated into the overall design of the shopfront.

The cornice is an important part of shopfront design, providing a horizontal line between the shopfront and the upper floors. Moulded cornices should have a detailed lead flashing for weather protection and, if the projection is sufficient the cornice may be used to incorporate a trough light if appropriate.

Pilasters and consoles relate directly to the classical architectural orders and, in general, their roles are decorative rather than functional. They make a significant contribution to the visual appearance of the shopfront by seeming to support either end of the fascia and so frame it. Pilasters and corbels should be used to provide vertical emphasis, give visual support to the fascia and upper floors and enclosure to the shopfront. And they establish physical separation from neighbouring buildings. They should be sufficient in size and number and appropriately located. They should project beyond the shopfront and be free of fixtures such as signs, alarm boxes and blind fittings.

These are traditional features and may not be appropriate for a modern shopfront design. However the principle of providing a clear division between shopfronts and visual support for the upper façade still applies. An appropriate way of achieving this in a modern design would be to incorporate or retain flanking masonry piers and where necessary, include uprights within the shopfront.

筑中使用。如在现代建筑上使用，需要采用嵌入的形式。

界定横带的装饰细节，如挑檐、遮盖匣和托臂等应该与店面的整体设计相协调。

挑檐在水平方向上连接店面和其所在的上层建筑结构，是店面设计一个重要的组成部分。模压制成的挑檐一般都会配备防雨板，起到保护挑檐的作用。如果突出深度足够大，则可以加入槽灯的设计。

相对于功能性和实用性，壁柱和支架的作用更倾向于观赏性，对于传统建筑来说，它们是一种必备的装饰元素。它们能够在外观上起到支撑横带两端，进而限定横带范围的作用。壁柱和支架通常为竖直走向，对横带和上层建筑结构起到视觉上的支撑作用，并且它们的宽度限定了店面的范围，使店面与相邻的建筑物很容易的分隔开来，不容易被混淆。壁柱和支架设计应当注意选择适合的尺寸、数量和位置，避开店面标识、报警信号箱和窗匣等装置。

这些传统装饰元素同样可能不适合现代店面装饰风格。然而现代风格店面中同样需要起到区分店面和视觉支撑作用的结构。如果需要在现代店面中使用这两类元素，则应注意对砌体结构进行整合或保留，必要时采用支柱。

Figure 6.7 Shoulder – Rio de Janeiro, Rio de Janeiro, Brazil, designed by Bob Neri, Carla Dutra, Fernanda Carvalho, photo by Marcos Bravo
Figure 6.8 XG Store at Renmin Road, Wenzhou, Wenzhou, China, designed by Guan Interior Design, photo by Fei Wang
图 6.7 里约热内卢 Shoulder 品牌店，巴西，里约热内卢，鲍勃·内里，卡拉·杜特拉，费尔南达·卡瓦略设计，马科斯·布拉沃摄影
图 6.8 XG（雪歌）温州人民路店，中国，温州，杭州观堂设计，王飞摄影

6.1.2 Signage Design

The primary function of a shopfront sign is to attract the attention of consumers and advertise essential information, for example the shop name, type of business, street number and if required the business address. It is important that signage is considered as an integral part of the design of a shopfront. Lettering, materials, size, colour, location and illumination all need to respect the character of the building and its surroundings.

Good and effective signs are simple, uncomplicated and uncluttered. If designed well signage can greatly enhance the appearance of the building and can also influence positively the trading success of a shop. The most effective signage is generally modest in scale, often just restricted to the entrance doorway. It is noticeable that the better quality the shop the smaller the signage.

The adoption of a corporate design approach may not be considered appropriate for every building or street. Where standardised treatment would dominate or have a negative impact on a building or street, it is expected that the design will be modified. Corporate styles are acceptable only where they are subordinate to architectural considerations.

Excessive advertising on the fascia should be avoided, as should unnecessary duplication of the shop name. Window stickers, poster displays and illuminated box signs in shop windows are often unsympathetic to the character of a building or area and would generally be discouraged. Window signs in upper floors will only be permitted for businesses operating solely on the upper floors. (See figure 6.4, 6.5, 6.6, 6.7, 6.8)

6.1.2 店面标识设计

店面标识的主要功能是吸引潜在消费者的注意，宣传店铺的基本信息，如店名、经营类型、街牌号码，需要时也可以包含地址信息。设计师应注意将标识作为店面的一部分进行规划设计，字体、材料、大小、颜色、位置和照明都应与所在建筑及区域相协调。

成功的标识设计通常具有简洁、易懂的特点，能够突出建筑特点，对店铺经营起到积极的推动作用。尺寸适中的标识最为实用，通常安放在店铺入口处。值得注意的是，越是高质量的店铺，选用的标识往往越小。

将企业标准设计用于店面标识并不一定适合每个建筑和街道。如果对建筑和街道景观构成消极的影响，则应该选择对其进行修改和调整。企业标准化设计方案需在符合建筑结构要求的条件下方可使用。

设计师在对店面进行设计时，应当避免在横带上出现过多的信息，不必要的重复店名也应当避免，过度的宣传往往会对宣传起到消极的作用。橱窗贴纸、海报展示和灯箱一般会与建筑风格不协调，应尽量予以避免。在店面上方楼层的窗口位置使用店铺标识仅限于宣传该楼层本身经营的业务。（图 6.4，图 6.5，图 6.6，图 6.7，图 6.8）

Figure 6.9 Home Cakery, Oporto, Portugal, designed by SPRS Arquitectura, photo by Rui Moreira Santos, Espinho
Figure 6.10 Kerry Center Concept Store, Shanghai, China, designed by Gruppo C14 srl, photo by Gruppo C14 srl

图 6.9 归巢甜品店，葡萄牙，波尔图，SPRS 建筑师事务所设计，鲁伊·莫雷拉摄影
图 6.10 嘉里中心概念店，中国，上海，Gruppo C14 设计公司设计，Gruppo C14 设计公司摄影

In some cases shop windows are filled with posters, stickers temporary banners and illuminated signs. This type of advertising can detract from the overall appearance of the shopfront and can detract from the appearance of the street, and will be discouraged. Window stickers should be restrained and cover no more than 20% of the total window area. Window signs in upper floors will only be permitted for a business operating solely on the upper floors. Lettering on windows using gilded paint in an appropriate style and size is very appropriate. On listed buildings this will normally require listed building consent and will generally be resisted.

6.1.3 Fascia Sign

Generally, one fascia sign per shop would be acceptable, as this is more effective than a number of individual signs all competing for attention.

Well-designed fascia signs can add decorative interest and project an image of quality, confidence and permanence. However, if poorly designed, they generate visual clutter and present an unattractive appearance. (See figure 6.9, 6.10, 6.11, 6.12)

As the fascia crowns the whole design of the shopfront, it should be in proportion to others nearby as well as the building on which it is located. Fascia signs should ideally be located at a minimum of 3 brick courses below the first floor windowsills and should not damage or cover any existing architectural feature. For example painted metal boxes fixed to fascias which project forward from cornice mouldings will not be acceptable. The latter suggestion is normally appropriate for public houses or restaurants where a pictorial sign is used, but it is

一些情况下，店面业主会要求设计师在设计橱窗时使用大量的海报、贴纸、临时横幅以及灯光装饰物等元素。这样的装饰方法会对店面形象乃至街道景观的整体效果形成干扰，并不建议采用。贴纸可以少量使用，它的面积应限制在整个橱窗面积的 20% 以内，在店面上方楼层的窗口位置使用贴纸仅限于宣传该楼层本身经营的业务。在橱窗上涂刷出文字是一个较为恰当的装饰手段。但在受保护建筑中使用需要预先获得许可，通常不会被批准。

6.1.3 横带标识设计

总体来说，每个横带对应一个店铺，同一个店铺使用独立的多个标识会分散注意力，降低宣传效率。

精心设计的横带标识可以强化装饰效果，向消费者传达高档、自信和持久可靠的信息。设计不当则可能会形成消极的店铺形象。（图 6.9，图 6.10，图 611，图 6.12）

由于横带位于整个店面的上端，设计师应注意其与相邻横带之间的大小关系，并且与所在建筑构成适合的比例。横带标识应位于二楼窗沿的下方，保持至少 3 个砖层的距离，能够避让开原先已有的建筑装饰元素。例如，与横带连接的，从挑檐延伸出的金属箱状物应避免使用。最后一条建议比较适用于公共建筑或餐馆等使用图形作为店面标识的场所，建议选择与横带平

Figure 6.11 Ligier, Buenos Aires, Argentina designed by Arquitectura y Diseño, photo by Julio Masri
Figure 6.12 Samba Swirl Frozen Yoghurt, London, UK, designed by Mizzi Studios, photo by Mizzi Studios

图6.11 利吉尔红酒店，阿根廷，布宜诺斯艾利斯，架构与设计工作室，胡里奥·马斯里摄影
图6.12 桑巴冰淇淋店，英国，伦敦，Mizzi设计工作室设计，Mizzi设计工作室摄影

often preferable that the sign is at fascia level.

Large scale detail drawings of proposed signs at a scale of 1:5 will be required for advertisement applications. Construction, lettering and graphic design must be clearly shown, showing cross sections, and with clear references to materials and colours.

The content should be kept to a minimum and contain only essential information. Telephone numbers and website addresses can be positioned in a less obtrusive place, such as on a door or in a window. Signs above fascia level will not normally be permitted.

Traditionally fascia signs were constructed of wood painted to a matt finish with sign-written letters, this should be the case when designing or restoring a traditional shopfront.

The use of effective lettering conveys an eye-catching and attractive image that can invite custom. But in considering the style of lettering, care must be taken to select an appropriate typeface, which not only reflects the character of the building but is also of the correct size and weight. The choice of lettering style used for the fascia sign is critical to achieving an attractive and enticing shopfront. The lettering and graphics on the sign should relate well to the nature of the business and the architectural style of the building.

The most appropriate location for lettering is on the fascia however there are some instances where lettering can successfully be incorporated in to the main window. They should be moderately sized and in proportion to the dimensions of the fascia. As a guide lettering should not be more then 65% of the height or 75% of the width of the fascia and should

行的位置。

申请时需要提供1:5比例的细节图。项目中的建筑、文字和平面设计方案需通过截面图清晰详尽地展示出来，对所用材料和颜色进行指示说明。

店面标识中的内容应当简洁、精炼。电话号码和网站信息要放在进门处或橱窗等次要位置。高于横带的标识通常是不被允许的。

传统的横带标识一般由木材涂刷亚光漆制成，上面绘制出店铺的品牌信息。设计师在对传统店面进行翻新设计时可以参考这种装饰手法。

标识上设计精美的文字信息能够吸引注意力，塑造有档次的店铺形象，起到招揽顾客的作用。对风格进行设定时，应注意选择适合的字体，使其不仅呈现出赏心悦目的外观，也能反映所在建筑的特点。为横带标识选择适合的文字风格对打造美观的店面设计至关重要。选用的文字和图样应与店铺经营的核心内容相关联，呼应所在建筑的建筑风格。

横带是整个店面最适合使用文字的位置，但也有一些情况下可以将品牌文字信息应用在橱窗上。这时应注意选择大小适度的字体，并与横带的规格成一定比例。字体内容不应超过横带高

Figure 6.13 Les Bébés Cupcakery, Taiwan, China, designed by J.C. Architecture, photo by Kevin Wu
Figure 6.14 Brokula&Z Experience Store, Zagreb, Croatia, designed by Brigada, photo by Domagoj Kunic
Figure 6.15 The House of Kipling, London, UK, designed by UXUS, photo by Dim Balsem

图6.13 贝贝杯子蛋糕店，中国，台湾，柏成设计，吴启民摄影
图6.14 Brokula&Z 体验馆，克罗地亚，萨格勒布，布里加达·格柏设计，朵马格·库尼奇摄影
图6.15 吉普林之家，英国，伦敦，UXUS公司设计，丁·鲍瑟姆摄影

be centrally placed. In general terms, the use of a lettering height should be 3/8 the depth of the fascia, with a maximum height of 300mm or less, subject to the scale of the building. Lettering and graphics should be clear and simple; and not dominate. Good effect can be gained by shading letters.

The content of signs should be kept to a minimum; any lettering and/or graphics should be in proportion to the dimensions of the fascia board. Oversized letters (in garish colours or materials) will not be supported and should be avoided, as should the repetition of a name on a single fascia. Transfer lettering may be a suitable alternative to hand painted lettering in some instances. In some instances individual cut out lettering to a painted timber fascia may be acceptable where the letters project no more than 10mm off the fascia. Plastic letters on historic buildings or buildings within a conservation area will rarely be acceptable.

Lettering on signs and fascias should enhance the appearance of the shopfront and the surroundings. The increased use of corporate styles has led to a loss of individual identity and harmed the character and appearance of many retail areas. There is a need to balance the requirements of national multiple retailers with a respect for the character of local areas. Standard house styles should be adapted to respect historic areas and buildings.

Where there is no proper shopfront, individual letters fixed directly to the wall without causing damage, or to window glass, can be used.

Simple brackets can look good, but depending upon location, a decorative wrought or cast iron bracket can be acceptable. Bracket design and fixings quality is important. Crudely

度的65%或宽度的75%，位于横带中央。总的说来，单个文字的高度一般为横带高度的八分之三，最高不超过300毫米，具体规格根据建筑的大小而定。文字和图样应清晰简明，不喧宾夺主。为文字添加阴影也能获得很好的效果。

标识的内容应当尽量简化，任何形式的文字和图样都应与横带构成适当的比例。应当避免使用面积过大，过于笨重（或者颜色和材料过于闪亮）的文字，或在同一块横带上过度重复出现店名。在通常情况下，如果文字的厚度不超过10毫米，在横带上安装制成文字比在横带上用颜料涂绘出文字更为适合。在受保护建筑或受保护区域内的建筑上使用塑料制成文字通常不被批准。

店铺标识和店面横带上的文字设计应该对店面和周围环境起到提升作用。过多使用标准化标识设计导致了多样化的缺失，使很多商业区失去了应有的个性与活力。商业区的整体设计需要平衡大型零售商和当地中小型店铺的不同需求。标准化设计方案应针对所在的建筑和区域进行调整。

如果没有正式的店面设计，可以选择在墙壁或窗玻璃上安装制成字母，作为店铺的标识使用。

支架单独使用时既美观又实用，但考虑到店面位置的区别，通常也会相应地配合锻造支架和铸铁支架来使用。支架的设计和

Figure 6.16 PCCW-HKT Signature Store, Hong Kong, China, designed by Clifton Leung, photo by Clifton Leung Design Workshop
Figure 6.17 Glassons Flagship Store, Broadway, USA, designed by Studio Gascoigne, photo by Patrick Reynolds

图6.16 电讯盈科旗舰店，中国，香港，梁显智设计，智设计工房摄影
图6.17 格拉森斯旗舰店，美国，百老汇，加斯科囚工作室设计，帕特里克·雷诺兹摄影

constructed mild steel brackets, which debase historic details, will not be acceptable, conversely there is considerable scope for carefully crafted contemporary brackets.

Colour is also important. Gilding or strong tones on a dark background reflect the light. Rich effects can be achieved by shading and blacking letters.

Multiples, Banks and Building Societies tend to have their own corporate identity and standardised signage. In sensitive locations, standard shopfronts can have the effect of diluting local distinctiveness. It is often possible to achieve a compromise so that the corporate image is maintained without eroding local character.

6.1.4 Hanging and Projecting Sign

Hanging signs are a traditional feature of shops and if well designed can add interest and originality to the frontage of the shop as well as help identify a shop from a distance, however they are not appropriate in for all shopfronts or in all locations. A proliferation of signs on all shops would create visual clutter therefore a balance needs to be reached and, in some cases, the architectural design of the building is so important that a projecting sign will be unacceptable. To ensure these signs and their supporting brackets should be carefully designed, relate to the size and scale of the building and be positioned to ensure that they do not damage or conceal architectural detailing. (See figure 6.13, 6.14, 6.15, 6.16, 6.17)

Only one hanging or projecting sign will be permitted per building. Signs are best placed in line with the level of the fascia and should not be above the level of the first floor sills. In

安装水准对于店面设计来说比较关键。受保护建筑上不可使用做工粗糙的钢架。

色彩也是设计中应该重视的问题。在深色背景中使用镀金、强烈的颜色会显得尤为明亮，使用阴影也可以增加浓厚的配色效果。

跨国公司、银行和建房协会通常有既定的企业形象和品牌标识，在一些敏感的区域，使用标准化店面设计会削弱当地风情和特色。设计师应注意对方案进行调整，既维护品牌的一贯形象，又尊重了当地的特色。

6.1.4 悬挂/突出式标识

悬挂式标识是传统店面的一个装饰特色，如果设计得当能够为店面增添不少趣味和创意，有利于消费者从远处进行辨识，吸引消费者的眼球。虽然悬挂式标识具有一定的积极意义，但它并不适合于所有的店面。如果一条街上的所有店铺都选择使用悬挂式标识，会使街道景观混乱无章。因此店面规划过程中应注意对标识与配套支架进行合理的设计，确保标识不会损坏或遮挡已有的或其他建筑装饰细节。（图6.13，图6.14，图615，图6.16，图6.17）

每个建筑只允许配置一处悬挂或突出式店面标识。位置应选在

Figure 6.18 Jocomomola Store, Madrid Spain, designed by churtichaga+quadra-salcedo architects, photo by Elena Almagro
图6.18 伊都锦服装店，西班牙，马德里，churtichaga+quadra-salcedo 建筑师事务所设计，埃琳娜·阿尔马格罗摄影

some instances it may be appropriate to site a hanging sign above the fascia level, however this will normally only be supported if it is to avoid damage to or covering up important architectural features. To avoid clutter only one sign will be allowed per shop.

Projecting signs at fascia level should be a maximum of 0.2 sqm, e.g. 500mm x 400mm. Hanging signs above fascia level where appropriate, should not exceed 800mm high by 600mm wide. As a general rule the bottom edge of the sign must be at least 2.6m above pavement level and the outer edge should not be within 1m of the kerb. A well-designed, traditional symbol representing the business can be an eye-catching alternative. On more modern buildings, simple projecting signs are effective.

Traditional style projecting or hanging signs on a decorative metal bracket can add interest to a building and the street scene. Where appropriate, these should be small and compact, made of wood or metal only and complement the business and shopfront. Where brackets already exist for hanging signs, these should be reused if they are of an appropriate design and suitable position. The design of new brackets should be appropriate to the shopfront and kept relatively simple.

Projecting box fascia signs are not appropriate on historic buildings and in conservation areas. If a box sign is to be used on a modern building, it should attempt to have a minimum impact on the design of the overall shopfront, ways of achieving this can include, being fully recessed behind the fascia, any lettering should be flush with the background panel or project only slightly and a matt finish can help improve aesthetic and visual appearance.

与横带平行的位置，不高于二楼窗沿。一些情况下，为了避免对其他结构的损坏或遮挡，可以将悬挂或突出式标识设置在高于横带的位置。同理，每个店面也只能使用一处悬挂或突出式标识。

横带高度的悬挂或突出式标识大小应不超过0.2平方米，如500毫米×400毫米规格。横带上方的悬挂或突出式标识高度不应多于800毫米，宽度不超过600毫米。通常，悬挂式标识的底端与人行道距离不应少于2.6米，标识外侧与路缘距离不少于1米。设计精良的传统式图样也可以作为悬挂式标识的一种替代，获得不错的宣传效果。

传统风格的悬挂或突出式店面标识能够为所在建筑和街道景观增添趣味。建议选择精巧、紧凑的设计，选用高质量的木材和金属，这样的成品对店铺的品牌形象和店面设计都会锦上添花。如果原有店面的标识已经配有支架，只要其设计合理，位置恰当，都可以选择继续使用。需要加入新的支架结构时，设计师应考虑到店面的风格，将其尽量简化。

突出式横带标识不适合在受保护建筑和地区内使用。如果应用在现代风格的建筑上，应尽量减少对店面整体设计的干扰，设计师可以考虑：将标识嵌入横带，文字内容应与背景板平齐或微微突出，亚光面也能传递出精致优美的感觉。

Figure 6.19, Figure 6.20 Filles du Calvaire, Paris, France, designed by Laurent Deroo Architecte, photo by C. Weiner _ LDA

图6.19，图6.20 受难修女街服装店，法国，巴黎，劳伦·德鲁建筑师事务所设计，C·维纳摄影

Often where the upper floor of a shop is in use by a separate business, there is a need for business nameplate located at the street entrance. The size of such plates should be of modest proportion and should not be illuminated.

In brief, projecting or hanging signs should not obscure architectural detailing. It should be located at fascia level and clear of the highway by a minimum height of 2.6m. From the carriageway, the minimum distance should be 1m.

6.2 Stall Riser Design

It should be noted that there would seldom be any situation where a stall riser is not required to complete the design of a frontage. It is often an integral design feature, providing support for the glazing and should therefore be a strong feature both in dimensions and structure.

The choice of depth will depend on the overall design of the shopfront and can be influenced by the depth of the fascia. Existing stall risers with decorative features of quality should be retained.

Materials used should always respect and enhance the materials of the whole building and shopfront. (See figure 6.18, 6.19, 6.20)

Contemporary designs should also incorporate some form of stall riser. These can be reinforced to provide additional security, can allow the display of goods at a more visible height and can help to create a horizontal link between adjoining buildings.

一些情况下，店铺只使用建筑的一楼，楼上的空间可能用于经营其他的业务。这时就需要在入口处使用铭牌进行标注指示。铭牌的尺寸以实用为主，不需要设计特殊照明。

简而言之，悬挂或突出式标识应在不遮挡其他建筑元素的前提下进行使用。高度与横带相当，离地超过2.6米，不阻挡行人通行。与车道相距1米以上。

6.2 竖板设计

需要注意的是，竖板是店面设计中不可或缺的组成部分。对橱窗玻璃起支撑作用，无论从尺寸还是结构上来看都是一个重要的结构元素。

设计师在对其进行设计时不能凭空想象，竖板的高度取决于整个店面的设计，也与横带高度有一定关系。如果店面原有竖板设计美观、具有特色，则应该尽量保留。

竖板使用的材料应与整个店面及建筑的选材相协调，追求和谐统一的效果。（图6.18，图6.19，图6.20）

现代的店面设计中也应该融合某种形式的竖板结构，除了能够提高安全系数，还有利于商品展示的进行，并与相邻建筑形成水平方向的联系。

Figure 6.21 Monki 3 Sea of Scallops, London, UK, designed by Electric Dreams, photo by Electric Dreams
Figure 6.22 Optic Shop Laskaris, Athens, Greece, designed by dARCHstudio, photo by Stathis Mamalakis
图 6.21 "扇贝之洋"服装店，英国，伦敦，电子梦想工作室设计，电子梦想工作室摄影
图 6.22 拉斯卡瑞斯眼镜店，希腊，雅典，dARCH工作室设计，斯塔西斯·马摩拉西斯摄影

6.3 Windows and Glazing Design

The size and style of shop windows, including mullions and transoms, should be in scale and proportion with the shopfront and the character of the building.

Their design should be dictated by the building's style. Windows should be taken down to a cill and stall riser. Large areas of undivided plate glass should be avoided as they give a blank aspect to the street and are expensive to replace. To overcome this, the window should be subdivided with vertical glazing bars known as mullions. Where glazing in shopfronts is to be divided, the number and location of the divisions should ideally reflect any existing vertical divisions on the upper floors. This will assist in providing visual support for the upper levels whilst providing a solid structural element at ground level. This principle is acceptable for modern shopfront designs, although subdivision is more commonly found on traditional shopfronts. (See figure 6.21, 6.22, 6.23, 6.24)

Windows should not be obscured by the proliferation of stickers or coloured film. Consideration must also be given to the display in windows. In some cases shop windows are filled with posters, stickers temporary banners and illuminated signs. This type of advertising can detract from the overall appearance of the shopfront and can detract from the appearance of the street. Laminated glass should be used for public safety, and as a security measure.

6.4 Doors and Access Design

Access to shops has to be given special consideration. Every opportunity must be taken to

6.3 橱窗设计

橱窗也是店面设计的一个重要元素，包括竖档和横梁在内，橱窗的大小和装饰风格应参考整个店面的大小、比例，以及所在建筑的风格。

这部分设计内容应与店面所处的建筑风格相一致。橱窗应与基石和竖板相连。在橱窗的设计上，应当避免大面积的使用无切割的平板玻璃，因为效果略显空洞，且一旦损坏更换起来费用昂贵。解决这一问题的方法是在竖直方向对玻璃进行分隔，使用到的玻璃柱叫做竖档。竖直方向的间隔分布应参考建筑内的其他楼层。这样能够在视觉上起到支撑稳定的作用。这类设计方法在传统店面中十分常见，同样也适用于现代风格的店面设计。（图6.21，图6.22，图6.23，图6.24）

橱窗上应当避免出现过多的贴纸和彩色贴膜。在橱窗处进行的展示要给予慎重的考虑。有时候店面橱窗会充斥海报、贴纸、临时横幅和灯光装饰。这种宣传手段会对整个店面外观乃至街道景观造成干扰，应注意避免。作为一种安保措施，也出于公共安全的考虑，建议橱窗使用夹层玻璃。

6.4 入口与通道设计

店铺的入口设计也是设计师应该关注的地方。条件允许的情况

Figure 6.23 NICKIE in Lishui, Lishui, China, designed by SAKO Architects, photo by Ruijing Photo
Figure 6.24 Sportalm Flagship Store Wien, Vienna, Austria, designed by BAAR-BAARENFELS ARCHITEKTEN, photo by Petr Vokal, Michael Alschner

图 6.23 丽水尼基儿童服装店，中国，丽水，迫庆一郎事务所设计，瑞金摄影
图 6.24 维也纳斯博塔旗舰店，奥地利，维也纳，巴尔·巴伦菲尔斯建筑师事务所设计，彼得·沃卡尔，迈克尔·阿尔施耐摄影

ensure that access to and circulation within shops is possible for all members of the public. All designs should conform to current standards of the Building Regulations where applicable.

Entrance doors should be clearly defined with a minimum clear opening of 800mm and 300mm clearance on the leading edge side. Handles should be easy to manipulate and clearly visible. Kicker plates should be fixed on the push side. Thresholds should be level and doors, preferably automatic should have minimum closing pressure.

Well designed entrance should follow below principles:
- Recessed
- Level Access
- Outward opening
- Easily distinguishable

Doors should be easy to open, or automatic. Revolving doors should be avoided but if essential should be supplemented by hinged or sliding doors that are available at all times.

Doors should preferably be located centrally to give visual interest and clearly define the entrance. The door must be easily distinguishable in the façade; recessing or using a detail colour can achieve this.

Doors should be partly glazed and clearly distinguishable. Glazed doors or large areas of glass should have visually contrasting areas in the form of a logo, sign or decorative feature at two levels in order to indicate their closed position.

下应该加强店内通行的便利程度，满足所有消费者的需求。设计方案应遵守当地的相关建筑条例。

入口大门应界限清晰，门口宽度不得低于800毫米，距离店面前端不少于300毫米。拉手应易于使用，安排在明显的位置。不锈钢踢板应安装在推的一侧。门槛要便于残疾人通行，门最好选用自动门，具有最小关闭压力。

设计合理的店铺入口需要符合以下要求：
- 嵌入式
- 便于残疾人通行
- 向外开
- 边界易于辨别

店铺大门应该方便人们开启，或为自动门。在店面设计中，通常不建议选用转门，但如必须使用转门，需设计相应的铰链门或滑动门加以辅助。

门的位置通常选在店面中央，外观与店面外观容易区分，这可以通过适当的凹陷设计或使用不同颜色来实现。

有玻璃的门比较容易分辨。店面中出现玻璃门或大面积玻璃的时候，应在其周围配有装饰性元素指示玻璃面的起始位置。

PART 6 Design Approach of Architectural Elements

Figure 6.25 How Fun Hair Salon, Taiwan, China, designed by JC Architecture, photo by Kyle You
Figure 6.26 Concept of How Fun Hair Salon, JC Architecture
图6.25 "多有趣"美发沙龙，中国，台湾，柏成设计有限公司设计，凯尔·尤摄影
图6.26 "多有趣"美发沙龙设计概念图，柏成设计有限公司

The design of the door should reflect a co-ordinated approach. Windows and doors should be made of the same material and painted the same colour. Fixtures and fittings should complement the style of the shopfront. Door panels should match the height of the stall riser. Attractive paving in the recessed entrance can enhance the character of the shopfront.

Creating independent access to upper floors, if they are in a different use, should be considered as part of any refurbishment scheme. The treatment of any such access should be in keeping with the materials and proportions of the shopfront.

Recessed doorways are a common feature of traditional shopfronts and add interest. They allow for an increased window area and a larger display. They can also be used to provide a level access by bridging any change in level between the shop floor and the street level. Where the provision of a recessed doorway is not possible it may be appropriate to install a ramp, either internally or externally. However, this would be dependent on the site and its surroundings and would need to be carefully and sympathetically designed and constructed.

Changes in level at the main entrance within shops should be avoided. If this cannot be done, a suitable ramp should be provided where possible. Integrated steps and ramps, ideally with a 1 in 20 gradient can sometimes offer a solution where there are steps into a shop. Where the design complements the existing building this can be preferable to a secondary side access, as this can be considered discriminatory. (See figure 6.25, 6.26, 6.27, 6.28)

Works within the footway usually require highway authority approval. As an alternative, with highway authority approval it might be possible to raise the height of the footway.

门的设计应该采用协调的方式进行。窗口和门应选用相同材料、相同颜色，辅助装置也应符合店面的整体风格。门板高度应与竖板相称。入口处路面的精致程度对提升整个店面设计水准有直接的影响。

店面翻修工程中，如果店铺楼上的空间属于其他商业用途，则应考虑为其设计独立出入口。独立出入口的设计应参考店面使用到的材料和建成比例。

嵌入式的门口设计是传统店面常见的一个结构，在设计上具有一定的个性。这样的设计可以增加橱窗的面积，创造更大的展示空间，还能够在店面和街道之间起到过渡作用。如果店面的条件不适合选择嵌入式门口，则可以考虑在室内或室外安装坡道。但是这对场地和环境也有一定的要求，需要进行仔细的勘查设计。

店面入口处的地面不应出现突然的高度变化，以避免地形的变化对消费者带来的不便，需要时可以设计一个坡道以方便消费者通行。倾斜度1：20的一体化台阶坡道是一个非常实用的设计方案，与单独的残疾人通道相比，一体化的坡道设计往往可以获得更人性化的效果。（图6.25，图6.26，图6.27，图6.28）

对人行道进行的工程改造应提前取得相关部门的批准。一些情况下，也有可能获准改变人行道的高度。

Figure 6.27 Aoyagi Souhonke Moriyama shop, Nagoya, Japan, designed by Yukio Hashimoto, photo by Nacasa & Partners Inc.
Figure6.28 Les Bébés Café & Bar, Taiwan, China, designed by JC Architecture, photo by Kyle You

图27 青柳森山店，日本，名古屋，桥本夕纪夫设计，Nacasa 摄影合作公司摄影
图28 贝贝咖啡馆，中国，台湾，JC 建筑设计公司设计，凯尔·尤摄影

6.5 Blinds and Canopies Design

Fixed blinds and blinds made from plastic or which have fluorescent or metallic finishes can detract from the appearance of many shopping streets. Dutch blinds and balloon canopies can look out of place too, whatever material they are made of and, will not normally be permitted.

Blinds and canopies are traditionally used to protect goods from damage by sunlight and give shoppers somewhere to shelter in bad weather. They also provide colour and interest, however, it is important that they are appropriate to the period of the building and the character of the locality, and are designed as an integral part of the shopfront. Where used, they should not detract from the style of the shopfront or from the character of the building or street scene.

Traditional retractable blinds were made of canvas, with a blind box incorporated into the fascia cornice. Where existing traditional blinds remain, we would encourage their retention and repair.

All canopies and blinds will be required to provide a vertical clearance of 2.3m from the footpath and a horizontal clearance of 1m from the kerb edge. A new blind or canopy should cover the width of the shopfront fascia and have the outer edge a minimum of 1m from the kerb and no less than 2.6m above the pavement. Highway regulations require that all blinds and canopies should be a minimum 2.4m above the footway and a minimum distance of 1m from the kerb. The design should be incorporated well into the overall shopfront design, flush

6.5 遮盖和遮阳篷设计

如今，很多店面都会选择使用遮盖或者遮阳篷，一是为了营造氛围，二是可以遮挡阳光。固定的遮盖，塑料或者具有反光效果的窗帘都可能对店面的外观造成不必要的干扰，从而影响消费者的视线。突出式遮阳篷无论使用何种材料，通常都不会被批准使用。

窗帘和遮阳篷既美观又可以遮挡商店里的货物，避免阳光对货物的曝晒，还可以为路上的行人提供遮风避雨的地点，给市民留下好的印象。遮阳篷的色彩和设计通常富于变化，但在设计的过程中也应当注意协调所在建筑和街道景观的风格和特色，并把它作为店面的一部分进行合理设计。

传统的可折叠式遮阳篷由帆布制成，配套的遮盖匣与横带挑檐相连接。如果在工程中遇到这种传统设计，建议设计师对其进行保留和修复。

遮篷和遮盖应在竖直方向与地面保持2.3米以上的距离，水平方向距离路缘1米以上。全新的遮盖或遮篷应与店面横带宽度相当，外缘与路缘距离不少于1米，下缘与地面距离不少于2.6米。公路相关法规通常要求所有的遮盖和遮篷距离路面2.4米以上，与路缘距离保持在1米以上。这部分设计应结合整个店面的效果出发，避免遮挡其他建筑装饰细节。遮阳篷的选材和

Figure 6.29 Burma Jewelery Boutique, Rue de la Paix., Paris, France, designed by Atelier du Pont, photo by Philippe Garcia
Figure 6.30 Fiori Flower Boutique, Kyiv, Ukraine, designed by Belenko Design, photo by Andrey Avdeenko
图 6.29 缅甸珠宝精品店，法国，巴黎，杜邦工作室设计，菲利普·加西亚摄影
图 6.30 菲奥里精品花店，乌克兰，基辅，别连科建筑设计公司设计，安德烈·阿夫迪恩科摄影

or behind the fascia, and not obscure any architectural detailing. The material and colour should complement the shopfront and building. They should not interfere with visibility of traffic signs or signals. If the blind covers all or part of an area where smoking occurs then it should comply with the relevant smoking legislation. (See figure 6.29, 6.30, 6.31)

Dutch blinds and similar non-retractable blinds are primarily used for advertising and are not traditional streetscape features.

New blinds should be of a traditional design in canvas or similar non-reflective material. Blinds will generally not be permitted above ground floor level or over doors and should usually cover the whole width of the shopfront fascia between the pilasters and be retractable into a blind box, preferably incorporated into the cornice, or fitted flush with the fascia. Care should be taken to ensure any blind does not obscure architectural features such as pilasters. Any lettering should be minimal and should not dominate the canopy area. The lettering style should co-ordinate with the design of the shopfront as a whole, especially the fascia sign.

Generally planning permission will be required to install a shop blind or canopy. If the canopy is to include advertising it may be that only Advertisement Consent is required. It is always best to check with the Local Planning Authority in the first instance.

Free-standing or fixed forecourt canopies require planning permission, and advertisement consent may also be required. Acceptable canopies are those which respect the character and architectural quality of the building and have limited impact on the street scene and the residential amenity of adjoining residents.

配色应当与店面和所在建筑的设计方案相一致。尤其应该注意的是，应避免对交通指示标识或信号构成干扰。如果人们可能在遮盖的位置范围内进行吸烟活动，则需要参考相关的条例和规定。（图 6.29，图 6.30，图 6.31）

突出式遮盖和类似的非折叠型设计一般用于宣传目的，并非传统的街道景观元素。

设计全新的遮盖结构时，建议选择使用帆布或其他非反光材料的传统式遮盖。一楼以上、门上等位置一般不建议使用遮盖。遮盖应以壁柱为界，宽度不小于横带，能够折叠收入窗匣。与挑檐相连，贴合横带更佳。应特别注意的是，遮盖不应遮挡壁柱等其他建筑结构。如果在遮盖上添加文字设计，应选择简洁的小面积方案。文字设计风格应与店面整体设计，特别是横带设计保持统一协调。

安装小型店面遮盖或遮篷通常需要预先获得当地部门的批准。如果遮篷上有广告信息，则只需要到广告管理部门提交相关申请。

使用独立遮篷和前庭式遮篷需要申请相关部门批准，可能还需要获得广告管理部门的许可。只有那些与建筑特点和风格协调，对街道和周围居民影响较小的遮篷才会得到批准。

Figure 6.31 Olivocarne Restaurant, London, UK, design by Pierluigi Piu, photo by Pierluigi Piu
图 6.31 橄榄树餐厅，英国，伦敦，皮耶路易吉·皮乌设计，皮耶路易吉·皮乌摄影

6.6 Display on Highway and Footpath

Displays placed upon the public highway can provide useful information to shoppers and enhance the character of the locality. Inappropriately placed displays can also hinder the passage of all highway users and in the worst cases present a serious hazard, particularly to the visually impaired or in narrow streets. The Local Highway Department treats all such displays as deposits upon relevant laws, has the right to remove any such deposits. Recognising the value that displays can add to the character of the locality, this action is only taken if the display is assessed to be an immediate or imminent hazard. Typically, if A-boards and displays are limited to one per frontage and not allowed to protrude more than 1.8m from the shopfrontage, and a minimum of 1.8m of clear footway remains, then the display will not cause a significant impediment. Please remember that the people passing your shopfrontage are all potential customers and that they have the right to pass and re-pass upon the public highway.

Displaying tables and chairs on footpath needs street trading licenses, which are applied for and regulated by the local government.

6.6 放置在公路和人行道上的宣传展示

在公路上进行宣传展示能够有效地向顾客传达商品和服务信息，强化街道特色。不合理的设计会影响到人们对公路的正常使用，甚至构成危险。公路管理部门对这类展示有相应的条文法规，有权对不合理的内容进行移除。但由于公路商品展示有增加城市个性的特点，通常只有展示物对公共安全构成直接危险时，相关部门才会采取行动。一般来讲，一个店面只能使用一块A字展板，且展板位置与店面距离不得超过1.8米，留出的通畅人行道宽度不得少于1.8米。这样才能避免对人流通行造成阻碍。设计师应意识到在店铺前方经过的行人都是潜在的顾客，应该为他们的自由通行提供方便。

放置在人行道上的桌椅需要获得经营许可，相关申请由当地政府发布和管理。

PART/7

第 7 篇 安保装置

SECURITY

It is acknowledged that security is a major consideration for shop owners and traders. Such security measures need to be considered in relation to the building and shopfront on which they are proposed to be located. Security measures by their very nature tend to be highly visible and therefore impact on the appearance of the building and the street scene. The installation of any security measures for a new shopfront should be considered at the design stage and not 'added on' as an afterthought. This will generally enable the shutter box and roller guides to be located behind the fascia and pilasters. Any security measures should be an integral part of the shopfront design and should endeavour to provide the least visually intrusive measures.

Planning permission is required for any external roller shutter and a consideration of the planning decision will be the visual appearance and impact that its installation will have on the character and appearance, not only of the building, but of the street scene. The security is necessary. However, it is important to recognise the need to achieve a balance between the need to address the security needs of any shop and the owners whilst responding to the wider environmental and public interests.

Their form and impact is such that when the shop is open the large external projecting box which houses the roller shutter mechanism is an unattractive fixture on the building and when the shop is closed and the shutters pulled down they create a visually 'dead' frontage. Many security measures can have a detrimental impact on the character or a streetscape and can create unwelcoming, 'dead' frontages. These intimidating 'dead' shutter frontages can have an intimidating effect on the street scene making, an area less attractive to visitors, especially during the evenings.

Shopfront security should be considered during the design stage and the physical solution should be restrained and unobtrusive. Careful forethought should be given to the siting, appearance and colour of security measures. Any application to install external shutters or grilles will be expected to evidence the crime history or future crime risk assessment for the property.

The use of laminated glass, internal brick bond style shutters and traditional stall risers to improve the security of shopfronts are supported.

Unobtrusive strengthening of the shopfront to protect against ram raiding is possible using elements of a traditional shopfront. It is relatively easy to provide a reinforced stall riser by

店面的安全问题无疑是店主和经营者最关心的一个问题。选择安保装置时需要结合店面所在的建筑以及店面的所处位置进行考虑。由于安保设备本身较为明显，多少会对建筑外观和街道的景观构成一定的干扰。新店面设计中应将相关设备考虑在内，合理美观地进行安置，可以将卷帘式壁龛和滚轮导轨隐藏在横带和壁柱之后。这样可以把安保系统对店面外观的不良影响降到最低。

在店面外部使用任何形式的安全卷帘都需要获得有关部门批准，其中主要的考量标准是装置对店面外观的影响，对店面所在建筑乃至街道的风格和特点可能造成的影响。安保措施无疑是十分必要的，但在安全需求与环境和谐之间达到平衡显然更为重要。

安保装置的形式对整个店面的设计有着至关重要的影响。以体积较大的卷帘匣为例，店铺营业时店主将卷帘收起，作为建筑外部结构，卷帘匣不甚美观；夜晚店铺关闭，卷帘展开拉下，店面看起来死气沉沉。许多安保装置都会给人留下这种冷漠呆板、缺乏活力的印象，这将削弱街景的美观度，降低店面对消费者的吸引力。

安保措施应当纳入店面设计的基本内容，并采取低调、保守的方法进行设计和处理。位置、外观和颜色是需要设计师着重考察的方面。是否安装外部护窗板和格栅应依据当地犯罪活动记录和风险评估决定。

设计师可以选择夹层安全玻璃、内部护窗板和传统的护窗挡板等方法提高店面的安全系数。

结合店面的其他基本元素，以相对低调的方式加强店面的安全保护比较具有可行性。相比之下，用混凝土和钢筋给店面外墙竖板加固相对简单易行。横梁和竖挡的部分也可以利用钢材进行加固。

一般情况下不建议采用以下的安保措施：

introducing concrete or steel behind the front façade. Steel can also be introduced behind transoms and mullions to provide additional strengthening.

The following security measures should be avoided and will not generally be supported:
• External shuttering of any kind;
• Solid shutters that prevent visibility into the shop outside operating hours;
• Horizontal slats that are at odds with a shopfront with a predominantly vertical emphasis;
• Security housing that break up the otherwise well-proportioned elements of the frontage.

7.1 External Shutters

Solid external shutters may often be preferred by shopkeepers, but they are visually intrusive, 'deaden' the street frontage and create an unwelcoming environment. They are vulnerable to graffiti and fly-posting. The need for shutter box housings and side runners also means that they can give the shop-front a bulky unattractive appearance. They are therefore the least acceptable form of security due to the visual impact on the host building and deadening effect on the street scene. External shutters are only acceptable in special circumstances with the support of relevant government or organization where there is a persistent problem of crime or vandalism which cannot be addressed by other measures.

Where the use of an external shutter is agreed, roller grilles or open weaved shutters are the preferred solution. The shutter box should be concealed within the fascia or installed flush beneath it. The shutter should be of a letter box style, allowing high visibility into the shop when down, and be coloured to match the shopfront. Uncoated or galvanised metal shutters are not acceptable. The guidance channels should be concealed or painted to match the shop frame or be removable during the day. Across recessed entrances hinged and demountable gates or brick bond style external roller shutters, where the coil can be concealed behind or inside the fascia, are acceptable. The architectural details of the shop-front must not be obscured or harmed by the fixtures. When the shutters are pulled down the pilasters should not be covered.

The installation of any external shutter will need to be fully justified and where it is deemed to be acceptable should only cover the glazed area of the shopfront and not the stall riser or pilasters. The housing for the roller shutter mechanism should be incorporated into the design of the shopfront so as not to be visible in the street scene. Externally mounted shutter housing boxes will not be acceptable

• 任何形式的外部护窗板
• 非营业时间对店内空间起到遮蔽效果的实体护窗板
• 在以竖直方向为主的店面上使用不和谐的水平板条装饰
• 破坏店面其他元素和谐度的安保装置

7.1 外部护窗板

尽管外部护窗板是许多店主心目中的安保首选，但由于它们外观不够美观，会使店面所处的街道景观变得呆板，缺乏亲切感。此外，护窗板还容易累积涂鸦和小招贴。

配套的板箱和轨道等也会使店面看起来笨重而有失美观。因此普遍将外部护窗板列为最不建议使用的安保措施。只有在特殊的情况下，当政府或相关部门认定犯罪、破坏活动无法通过其他方法来解决，才能选用外部护窗板。
获批使用外部护窗板后，应优先选择格栅卷帘，将卷帘匣隐藏于横带之中，或安置在横带下方，与横带齐平。拉下这样的护窗板时，人们仍可以清楚地看到店内的情况。护窗板的颜色可以选择，应尽量配合店面整体配色。

护窗装置应避免使用裸露金属或镀锌金属。固定护窗板的凹槽或轨道应该选择隐藏或可拆式设计，或涂刷成适宜的颜色，保证白天的营业时间内不会对店面外观构成影响。护窗板应避让店面的其他建筑装饰元素，使用时不应遮挡店面的壁柱结构。

安装任何形式的外部护窗板都需要进行充分的论证，即便获准安装，也只能覆盖橱窗的玻璃部分，而不应盖住竖板或壁柱等其他结构。设计师应将护窗板的收纳装置融入整个店面设计，尽量隐藏在其他结构之中。

7.2 内部护窗板和格栅

店面设计中也可以在橱窗内部使用轻型格栅或网格式护窗板。这种结构便于清洁，起到一定保护作用的同时使橱窗呈现通透开阔的感觉。

7.2 Internal Shutters and Grilles

These consist of light mesh grille or lattice roller shutters and can be fitted discreetly behind the shop window. They are easy to keep clean and in working order because they are not exposed to the weather. They allow the window to retain an 'open' appearance but maintain a high degree of security for the goods.

An alternative to external roller shutters, as these are not always appropriate, may be the installation of internal shutters on a building. Ideally these should be set back from the display window as far as possible and painted in a colour to co-ordinate or compliment the shopfront.

Internal lattice or brick bond roller type grilles can be set between the display and the glass. The coil can be fitted in an existing false ceiling or the window soffit and not seen from outside.

These should be perforated and designed to be in keeping with the design of the shopfront.

7.3 External Grilles

Grilles are fixed to the outside of windows and doors on runners or on hooks and padlocked to the window frame. They also give security while maintaining an open appearance. Their physical impact is minimal because they do not require any box housings or side rails. The grilles should be removed during hours of business and stored internally. They should be lightweight and not damage any architectural features.

As an alternative to solid roller shutters, external demountable mesh grilles painted in a dark colour and placed over windows, can be supported. Shutter guides should be removable or integrated into the pilasters or glazing bars and painted to match.

If external grilles are proposed, the following criteria should be observed:
• The window display should still be easily visible;
• Any associated mechanisms and housing should be concealed behind the external structure (i.e. behind the fascia or recessed and flush with the shopfront).
• The fascia, stall riser and pilasters should not be covered by the grille;
• Runners or shutter guides should be either removable or integrated into the pilasters or

内部护窗板为外部卷帘提供了很好的补充和替代，安装的位置应尽量远离橱窗玻璃，色彩选择应与店面外观相协调。

内部格栅可以出现在展示商品和玻璃之间。配套的线圈可以连接在已有的吊顶上，也可以连在窗框下端。这样的连接处从店面外部是观察不到的。

相关设计细节也应与店面整体设计方案相协调。

7.3 外部格栅

通过滑道、挂钩或锁扣安装在橱窗和门外侧的格栅可以兼顾店面的安保和视觉开阔性。由于格栅不需要配套的收纳装置和栏杆，对店面外观的影响很小。营业时间内将格子窗摘除，收于店内，因此需要采用轻质材料，避让其他建筑结构。

网状格栅可以作为实体卷帘的替代物，涂刷成深色，应用在橱窗上，起一定的保护作用。建议选择隐蔽或可拆式的导槽，或将导槽与壁柱整合在一起，涂刷成相应的颜色。

使用外部格栅时应注意以下几点：
・格栅不影响橱窗展示
・将相关装置隐藏在横带等其他外部结构之中
・避开横带、竖板和壁柱
・选择隐蔽或可拆式的导槽，或将导槽与壁柱整合在一起，涂刷成相应的颜色
・格栅的颜色也应与店面整体搭配
・格栅本身及其配套装置应该避让店面原有的建筑装饰元素

7.4 夹层安全玻璃

另外一个提高安全系数的方法是在橱窗上使用夹层安全玻璃。尽管在受保护建筑中可能不被批准，但夹层安全玻璃可以广泛应用在多数店面中，减少犯罪行为的发生。

glazing bars and painted to match;
• The grille itself should be painted in a colour appropriate to the rest of the shopfront;
• No part of the grille or its housing should obscure or damage any important architectural features present on the shopfront.

7.4 Laminated Glass

Another alternative to the installation of shutters is to use laminated glass for the shop windows. Although this may not be acceptable on listed buildings within other shopfronts this type of glazing could be incorporated into the design and may help to reduce crime.

Laminated (Security) Glass has the capacity to remain intact even when broken. Laminated glass offers protection without adversely affecting the appearance of the shopfront as no additional or fixings are required. Laminated glass should therefore be the first solution to be considered.

Toughened Glass or architectural Perspex is similar alternative.

7.5 Other Measures

Reducing the size of windowpanes can provide less of a temptation to vandals and reduces the cost of replacing glass, but the suitability of this will depend on the design of the host building.

Alternatively, removable security screens can be fixed to the window area of a shopfront outside normal trading areas. There are several examples of such screens in the County, which have been designed attractively in materials and finishes that do not detract from their surroundings.

A common security feature is the use of fire and burglar alarms and while often essential these can often be unattractive and obtrusive if sited incorrectly. Alarms are best incorporated on centre lines between upper windows or within recessed doorways. Burglar alarms and telecom junction boxes and any other wiring should not conceal architectural features or be located in over-conspicuous positions. Alarms should never be sited on architectural features such as corbels or pilasters. Any wiring should be neatly fixed or hidden. Non-ferrous fixings will avoid problems with rust. Consent for these may be required on a listed building.

夹层安全玻璃受到击打时也能保持原样。由于无需额外的固定装置，它既能提供安全保护，又基本不会改变店面外观。因此夹层安全玻璃是安保装置中的优先选项。

钢化玻璃和有机建筑玻璃与其属性相近，建议优先使用。

7.5 其他安保措施

减小橱窗玻璃的面积可以抑制破坏行为的发生，降低更换玻璃的费用。适合与否取决于店面所在建筑的设计风格。

也可以选择安装在橱窗外的移动式保护屏。恰当的选材和配色可以使其做到与周围环境和谐一致。

店面中还有一个常见的安全装置：火灾和防盗警报器。这些装置几乎每个店面都有，但设计、安装不当也会对店面外观造成不好的影响。警报器最好安装在上层窗户的中心线上，或者安装在嵌入式门口内侧。防盗警报、电信分线盒和其他布线应避开原有建筑结构，选择不显眼的位置。警报器要避开托臂和壁柱，布线应该固定整齐或隐藏起来。使用有色金属配件可以避免零件生锈的问题。在受保护建筑使用以上装置可能需要申请批准。

PART/ 8

第 8 篇　设计案例

Puma Store London

伦敦彪马旗舰店

Design: Pajer & Franz Studio
Location: London, UK
Completion: 2012
Photography: Manuel Schlüter / Copyright Puma AG

店面设计：Plajer&Franz 工作室
店铺地点：英国，伦敦
竣工时间：2012 年
图片摄影：曼纽尔·施吕特尔 / 版权归彪马品牌所有

The factor 'fun' is reflected in various regional elements, which gives the store its local relevance. This begins with the façade covered in 3D images of the traditional London red cell booth and continues inside the store, where the 'pumarized' cell booth reappears serving as display system and dumb waiter cladding.

The new puma store front design in London's Carnaby street elevates puma's effort in bringing back the joy of sport and lifestyle into retail environments. The new store front surprises with unexpected features, reflecting the passion in interacting with consumers, and creating a memorable and engaging retail experience with local flavour.

店面设计中的"趣味"主题通过多种地域元素表现出来。外墙上以伦敦红色电话亭为主的传统立体图案别出心裁，与室内装饰元素相呼应。

伦敦卡尔纳比街的彪马新店重新将快乐运动和生活的理念引入了店面环境。新店面充满新奇意外的惊喜，体现品牌与消费者沟通、互动的热情，营造一个难忘的消费体验。

Como quieres que te quiera...

"想你所想……"服装店

Design: Arquitectura y Diseño
Location: Buenos Aires, Argentina
Completion: 2012
Photography: Julio Masri

店面设计：构架与设计工作室
店铺地点：阿根廷，布宜诺斯艾利斯
竣工时间：2012 年
图片摄影：胡里奥·马斯里

The main idea of this brand in their shops is to recreate the bedroom of a teenage dream, creating a surrounding of female and classical romance. The storefront reconstructs the image of a typical old house in Buenos Aires, for its classical forms and details. It is built from solid wood. The door is highlight of the project, as it reused the original door of an old house built in 1928 to achieve the image of memories. The purple, flowery patterns and stained glass are the key elements to attract teenage girls, romantic and elegant clients who are the target clients of this brand.

本项目的主要设计目标是重建一个少女心目中的理想卧室，一个充满女性化的浪漫气息和经典元素的购物环境。店面重现了布宜诺斯艾利斯典型的老房子形象，包括其经典的形式和细节，使用了大量实木。作为工程中的亮点，设计师将一座1928年的老房子里拆下的门使用在这里，创造出充满回忆的感觉。绚丽的紫色图案和彩色玻璃对于吸引少女以及追求浪漫和优雅的目标消费者十分有效。

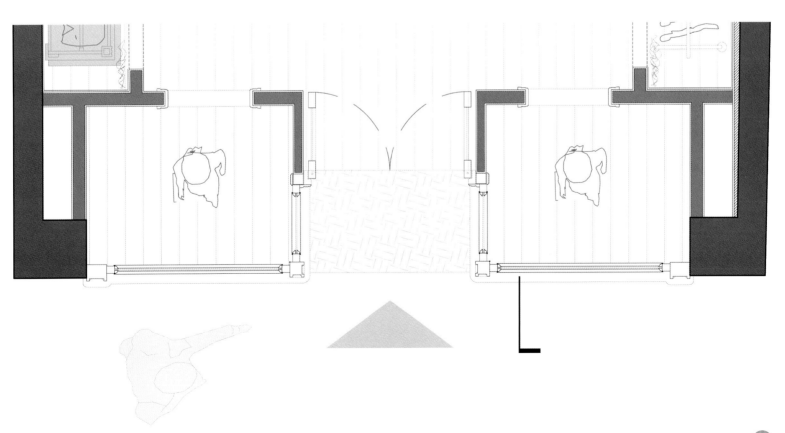

Farm - Niterói

尼泰罗伊 FARM 精品服饰店

Design: be.bo. /Bel Lobo, Patricia Batista, Alice Tepedino, Ana Luiza Neri, Ayla Carvalhaes, Clarisse Palmeira, Fernanda Mota
Location: Niterói, Rio de Janeiro, Brazil
Completion: 2012
Photography: Marcos Bravo

店面设计：be.bo. 设计公司 / 贝尔·洛沃，帕特里夏·巴蒂斯塔，艾里斯·特贝蒂诺，安娜·路易莎·内里，艾拉·卡瓦海斯，克拉丽丝·帕尔梅拉，费尔南达·莫塔
店铺地点：巴西，里约热内卢，尼泰罗伊
竣工时间：2012 年
图片摄影：马科斯·布拉沃摄影

In 2009, when Farm's first flagship store was designed, be.bo. developed a white iron mesh inspired by a traditional wicker craft work to reduce the impact of direct sunlight on its North façade. Since then, this white mesh has been used on every new store be.bo. designed for this brand, reinforcing its visual identity, although it appears in a different shape in each store.

For this Niteroi store, the designers converted the former building's entrance, narrow and with low ceiling height, into a gallery which takes the customer to the actual store.

2009 年，FARM 服饰品牌的首家旗舰店由 be.bo. 设计公司精心打造。设计师以传统柳条手工制品为灵感，创造了一个白色铁网结构，减少直射日光对北外墙的影响。从那时开始，白色网状结构就被应用在 FARM 品牌所有的店面设计中。尽管在各个店铺内白色网状元素以不同的形状出现，其独特的形式对品牌形象起到了有效的强化作用。

在尼泰罗伊的这个新店项目中，设计师将建筑原来狭窄、低矮的入口改造成了极具吸引力的艺术通道。

Guru Palermo

巴勒莫大师品牌旗舰店

Design: DUCCIO GRASSI ARCHITECTS
Location: Palermo, Italy
Completion: 2010
Photography: Andrea Martiradonna

店面设计：杜乔·格拉西建筑师事务所
店铺地点：意大利，巴勒莫
竣工时间：2010 年
图片摄影：安德里亚·马蒂拉多纳

The Guru space in Palermo sets in the 'noble' architecture of Viale della Libertà with the spontaneity, the energy and the life mark typical of the Guru brand.
With no intention of being disrespectful to the building architecture, the atmosphere, the spaces and the unusual and out-of-context materials highlight the Guru lifestyle, marked by irony, mainly self-irony, paradoxes, open matching and free expression.
The diversity of the Guru shop in Palermo is evident both from the exterior, where there are no windows, and the access, reacheable through a garden at a raised floor. On the exterior, in front of the stairs, a giant Guru sign lays on a real meadow, vertically placed. The outside flooring is made by pre-pressed concrete.

巴勒莫的大师品牌旗舰店坐落在"高贵的" Viale della Libertà 建筑之内，店面设计突出品牌随性而充满能量的特点。
大师品牌推崇的生活方式充满讽刺、自嘲、矛盾、开放性和表述的自由。针对这一特点，设计师对建筑原本的风格和设计表现出充分的尊重，对其中一些不常见、不容易被接受的装饰元素给予了全面的考量。
无窗的外墙和入口设计彰显店铺的独特个性。外部楼梯前面，巨大的 GURU 标识矗立在草坪上，室外地板采用的是预压混凝土材料。

Poroscape

Poroscape 品牌时装店

Design: Younghan Chung
Location: Jongrogu, Seoul, Korea
Completion: 2011
Photography: Kim Jae Kyeong, Park Jong Min

店面设计：郑荣昊
店铺地点：韩国，首尔，钟路区
竣工时间：2011 年
图片摄影：金栽经，朴中民

This author selected brick among the materials already having the porosity of a physical property itself with the aim of expressing the process of forming the weft and warp of fabric and its shape metaphorically. The brick-laying method connoting cumulation requires temporality and precision at the same time, and physical properties of brick is slow in their change for a long time even after the completion of brick-laying process. The project team decided on a brick for reason of some attributes of such physical properties.
In addition, they proposed some stacking methods. Generally, the brick-laying method has a limit to the independently supportable height in cumulating bricks, so they also needed the detailed structural frame as a supplementary means. There were an unrealistic alternative and also the option to overcome the limit to vertical height out of some stacking methods.

设计师选择了砖石作为外墙设计的主要材质，砌砖对速度和精度都有一定的要求。砖石的外观也会随着时间发生一定的变化，这也是设计师选择它的一个原因。
此外，设计团队还针对砖墙提出了一些策略。使用传统砌砖方法的独立结构在高度上有一定的限制，需要详细的结构框架。新方法有助于克服这种限制。

Givenchy Flagship Store Seoul

纪梵希首尔旗舰店

Design: PIUARCH / Francesco Fresa, Germán Fuenmayor, Gino Garbellini, Monica Tricario
Location: South Korea, Seoul
Completion: 2014
Photography: Kyungsub Shin
Used materials: Oak Wood, Calacatta Marble, Sala Noir Marble, Electropolished Stainless Steel moulded, Black Basalt Marble

店面设计：PIUARCH 设计公司 / 弗朗西斯科·弗雷萨，热尔曼·富恩马约尔，基诺·加博里尼，莫妮卡·特里卡里奥
店铺地点：韩国，首尔
竣工时间：2014 年
图片摄影：申景夑
使用材料：橡木，卡拉加塔（鱼肚白）大理石，黑色萨拉大理石，电解法抛光不锈钢，黑色玄武岩

The Seoul flagship store is located in the blossoming shopping district of Seoul, Gangdam in Cheongdam Dong. The concept guidelines provided by the client have been developed introducing a number of innovations such as the use of natural materials - marble and basalt stone in particular - new furnishings typology, new articulations of space. This intervention creates a new, refreshing environment, according with the image of the fashion brand. The concept of the facade - designed as a second skin, identity of urban expression – is the evocation of the distinctive tailoring 'T CUT' that characterizes the style of the French brand. The embossed metal sheet used for the façade surface, creates an effect that changes depending on the reflection of light and the surrounding landscape. The design has being inspired by the Optical Art as well as the processing techniques of fabrics used by fashion house.

纪梵希首尔旗舰店位于首尔的清潭洞时尚购物区。委托方提供的设计概念中引入了一些创新元素，例如大理石和玄武岩等天然建筑材料的使用、新型家具的采用、新的空间连通设计。这些设计形式打造出一个让人耳目一新的环境，彰显了品牌的形象与个性。外墙设计呼应纪梵希品牌独特的 T 形剪裁，使用的压花金属板产生随光线和周围环境变化的奇特效果。视觉艺术以及服装制作过程中对面料采用的处理技术是这一店面设计的主要灵感来源。

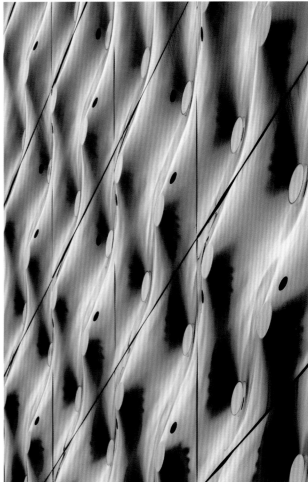

PART 8 Works-Fashion 087

Salinas – São Paulo

圣保罗萨利纳斯旗舰店

Design: Bel Lobo, Mariana Travassos, Patricia Fontaine, Patricia Gava
Location: São Paulo, Brazil
Completion: 2011
Photography: Graziella Widman

店面设计：贝尔·洛沃，玛丽安娜·特拉瓦索斯，帕特里夏·方丹，帕特里夏·加瓦
店铺地点：巴西，圣保罗
竣工时间：2011 年
图片摄影：格拉茨·威德曼

Salinas is a swimwear brand, very representative of Rio´s tropical vibe, that was opening its first flagship store in São Paulo. On this façade design the designers intended to keep the fresh and colourful identity of the brand, while using a material more related to the urban context in which this store was inserted. They chose corrugated metal sheets painted on pink because it also evoked the vertical wood paneling used on other Salinas stores. The store's access doors refer to typical brazilian beach houses, and provide an interesting contrast with the corrugated metal sheet.

萨利纳斯是巴西的泳装品牌，经营内容具有典型的里约式热带风情。本项目是品牌在圣保罗的第一家旗舰店。店面外墙设计使用的是与圣保罗城市景观相称的建筑材料，同时也保留了清新、明艳的品牌特色。粉色的波纹金属板呼应萨利纳斯品牌的其他店面。店铺大门象征巴西的沙滩小屋，与波纹金属板形成有趣的反差对比。

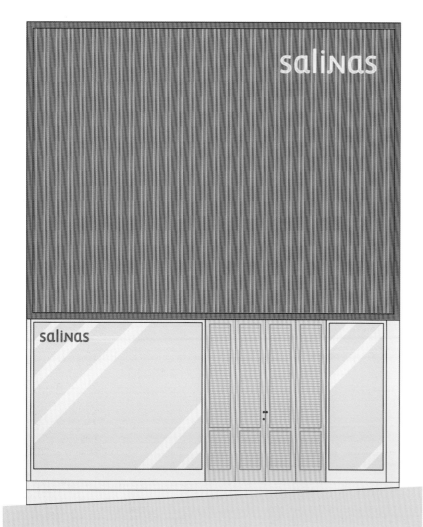

Shine Fashion Store

Shine 时尚精品店

Design: Nelson Chow (NCDA)
Location: Hong Kong, China
Completion: 2011
Photography: Dennis Lo Designs

店面设计：NCDA 建筑设计有限公司
店铺地点：中国，香港
竣工时间：2011 年
图片摄影：丹尼斯·卢设计公司

Shine is one of Hong Kong's most renowned high end multi-brand fashion stores, known for bringing pioneering foreign brands to the trend conscious locals.

Inspired by the name of the store, a 7m tall asymmetrical glowing star-like structure forms the primary street identity along Leighton Road, attracting both pedestrians and motorists. The pristine white shell embodies a black interior wall that further unfolds to create three main rooms: The entrance gallery, the upper level sales area and finally the dressing room. The design of the Shine flagship store in the Leighton Center showcases how the idea of a 'shining star' could be translated architecturally into a fashion retail space, creating a visually striking yet highly functional contemporary store.

作为一家香港著名高级流行服饰总汇店，Shine 一向以独到眼光及敏锐触觉为时尚名人搜罗环球顶尖品牌服饰。

铜锣湾礼顿中心新店的室内设计灵感是来自 Shine "闪亮"的概念，突显 Shine 的前卫和独特的个性。正门是高 7 米、呈现不对称的星形外貌，成功吸引礼顿道上的行人和驾驶者的目光。正门的纯白色反衬室内的黑色墙壁，进一步展示店内的三个间隔空间：正门入口、楼梯之上的两个陈列区及更衣室范围。

从正门步入店内，设有三个陈列位置和悬挂于楼梯两旁的人形模特儿，是展示时装买手每季搜罗到的精选货品的舞台。街上行人经过店铺时也会被这独特的视觉效果吸引。

Shine 的礼顿中心旗舰店重新演绎不规则星形的建筑理念，并将这几何概念完全融入到时装零售的空间之中，成为一家外观突出、功能齐全的现代时装店。

Shoulder – Rio de Janeiro

里约热内卢 Shoulder 品牌店

Design: Bob Neri, Carla Dutra, Fernanda Carvalho
Location: Rio de Janeiro, Brazil
Completion: 2012
Photography: Marcos Bravo

店面设计：鲍勃·内里，卡拉·杜特拉，费尔南达·卡瓦略
店铺地点：巴西，里约热内卢
竣工时间：2012 年
图片摄影：马科斯·布拉沃摄影

Located on Ipanema's main street, Shoulder's façade is covered with horizontal wood venician blinds without any unnecessary decoration, inspired on 'brise-soleil', a variety of permanent sun-shading structures, ranging from the simple patterned concrete walls, which is a very representative element of Brazilian modern architecture. For the exterior skin hard wood was chosen because it is a natural material and yet very sophisticated, according to Shoulder's branding: Natural, casual, free. The brand sign is installed at the top and on the shop window, echoing each other. Seven exquisite spot lights with the simple lines are employed, which is suitable for the succinct sign of the brand.

店面位于依帕内玛主街之上，外墙采用水平的木质百叶窗造型，摒弃了不必要的装饰细节。这一设计灵感来自巴西现代风格建筑中较为代表性的建筑元素，一种永久性的百叶窗遮阳结构，通常与简单的混凝土墙组合使用。外墙表面使用的是硬木材料，天然环保又具备精致的格调，它的特质可以用品牌形象来表达：天然、休闲、自由。品牌标识分别安装在外墙上端和橱窗之上，二者相互呼应。设计师在七处位置安装了同样风格简洁的射灯，配合简明的品牌标识。

Deskontalia Shop

Deskontalia 店铺

Design: VAUMM Architecture & Urban Planning
Location: San Sebastian, Spain
Completion: 2012
Photography: Aitor Ortiz

店面设计：VAUMM 建筑和城市规划公司
店铺地点：西班牙，圣塞巴斯蒂安
竣工时间：2012 年
图片摄影：埃托尔·奥尔蒂斯

For Deskontalia store, located in an urban downtown street, the space should become not only a space to sell, but a space to be a meeting point between brand and people, an open space, a place of the city where an online business becomes a physical reality. The space has been treated as a white empty space where old items such as masonry walls or casting pillars are bathed in this colour, as well as more contemporary new resin pavement, in an attempt to transform the store not in a shop but in a store where different transformations may occur.

坐落在城市商业区的商店不仅是出售商品的空间，也是品牌与大众相遇的地点，是一处公共空间，是网络交易在城市中的一个落脚点。设计师将店内原有的砖墙和支柱等结构重新粉刷成白色，新增的现代风格树脂路面也选择了白色，将店面空间还原为一个空白的空间。目的是将店面改造成一个超越店铺本身的，具有多种可能的终极空间。

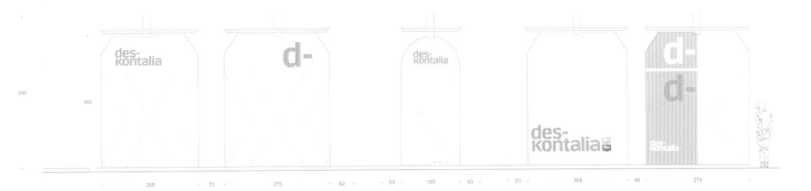

Plac Jeans Flagship Store at Myung-dong

PLAC 牛仔明洞旗舰店

Design: URBANTAINER Co., Ltd.
Location: Seoul, South Korea
Completion: 2010
Photography: URBANTAINER

店面设计：Urbantainer 设计有限公司
店铺地点：韩国，首尔，中区
竣工时间：2010 年
图片摄影：Urbantainer 设计公司

PLAC Jeans is the premium denim brand based in Korea offering more than 10 denim fittings and styles. The designers' goal is to engage the consumers who visit the retail space to gain an easy access with various approaches to find the perfect denim that best fits their preference and body types. Façade of the building made of contorted and intense orange glass box make a fun and impressive entrance.

PLAC 牛仔是韩国的一家高端牛仔休闲服装品牌，提供十余种不同的版型和风格。设计师试图让进店的消费者能利用设计好的多条路线轻松找到适合自己的服装产品。建筑外墙扭曲的形式和内外呼应的橙色玻璃盒形象使店铺个性分明，让人过目不忘。

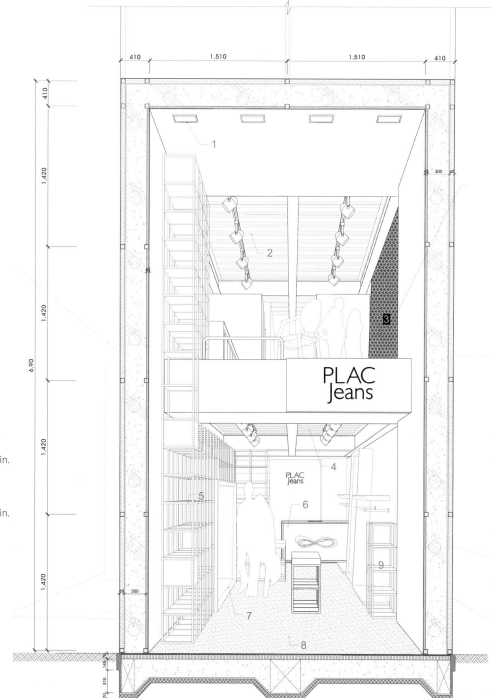

1. 10 in HQI 25ow multi
   Size:W 275mm×d130mm ×H 150mm
2. App. Grey water paint fin.
3. T 8mm mirror glass fin.
4. App. grey water paint fin.
5. 20×20 G/V steel plate / app. white emulsion paint fin.
6. T 8mm tempered glass/ prange veneer sheet fin.
7. T 18mm mdf/ app. prange emulsion fin.
8. App. selfleveling fin.
9. 20×20 G/V steel plate / app. white emulsion paint fin.

1. 25 瓦金属卤素灯
   尺寸：宽 275 毫米 × 深度 130 毫米 × 高度 150 毫米
2. 指定灰色水性涂料表面
3. T 8 毫米反光镜面
4. 指定灰色水性涂料表面
5. 20×20 G/V 钢板 / 指定白色乳胶漆表面
6. T 8 毫米钢化玻璃 / 层压板表面
7. T 18 毫米中纤板 / 指定乳胶漆表面
8. 指定自动调平装置
9. 20×20 G/V 钢板 / 指定白色乳胶漆表面

Jolie VT

Jolie VT 服装店

Design: Arq. Mário Wilson Costa Filho
Location: Fortaleza, CE, Brazil
Completion: 2012
Photography: Gentil Barreira

店面设计：马里奥·威尔森
店铺地点：巴西，福塔雷萨
竣工时间：2012年
图片摄影：让蒂尔·巴雷拉

The idea of the façade is to do a reference to the architecture of industrial buildings typical of Soho in Manhattan, but with a new look.
Imposing façade marked by its big height and only two modern and simple finishes: the red brick in contrast with the black colour. The elevation follows a symmetry of right angles and the glass wall of windows gives high transparency which allows the interior of the store to be seen from afar.

During the night, the lighting is another important feature to highlight the volumetric façade and interior.

这个服装店的外墙设计灵感源自曼哈顿 SOHO 区典型的工业建筑设计，但在此基础上进行了一些改进。
高大的店面外墙使用了红砖和黑色的对比，简洁而现代。对称的立面设计与玻璃橱窗保证了较高的透明度，使人们在与店面有一定距离时就可以看到店内的情况。
夜间，照明也是加强店面立体感的一个重要手段。

Cream

克里姆女装店

Design: Rodriguez Studio Architecture p.c
Location: New York, USA
Completion: 2008
Photography: JC Paz

店面设计：罗德里格斯建筑设计工作室
店铺地点：美国，纽约
竣工时间：2008 年
图片摄影：JC·帕斯

Cream is a boutique on Manhattan's Upper East Side in New York City. The storefront design concept draws from jewelry window displays to focus attention on precious objects.

From blocks away, the bright porcelain tile façade stands out from its monotonous surroundings, creating a sense of oasis. In the evening, the lyrical signage emits a subtle neon glow. Upon closer inspection, the strong, boldly coloured framing of the display window- inspired by jewelry display windows- highlights the featured apparel and invites entry.

克里姆女装店是纽约曼哈顿上东区的一家精品服装店，店面通过珠宝橱窗般的设计展示并吸引人们关注精致的服装产品。与周围的单调环境相比，克里姆女装店明亮的瓷砖外墙脱颖而出，从几个街区以外就可以看到，好像沙漠中令人心驰神往的绿洲。夜幕降临时，店铺标识散发出柔和的氛光。橱窗边框的设计灵感来自珠宝展示橱窗，配色强烈、大胆，突出主打服饰，吸引消费者进店。

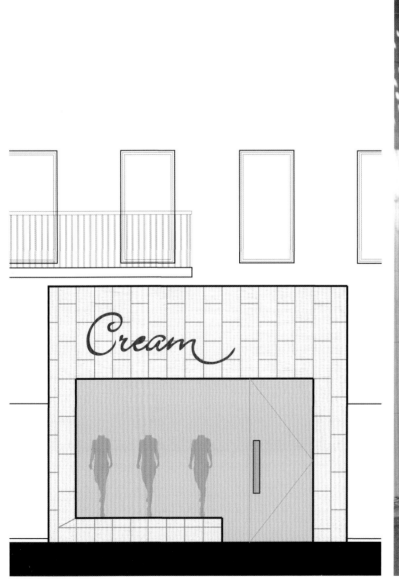

Brokula&Z Experience Store

Brokula&Z 体验馆

Design: Brigada, Damjan Geber
Location: Zagreb, Croatia
Completion: 2012
Photography: Domagoj Kunic

店面设计：布里加达·格柏，达米扬·格柏
店铺地点：克罗地亚，萨格勒布
竣工时间：2012 年
图片摄影：朵马格·库尼奇

The concept of the flagship Brokula&Z mono-brand store in the center of Zagreb, consistently follows the ecological foundation of the brand. By pushing back the shop window to create a ledge topped by a plush, branded cushion, the designer created an inviting public seating that relaxes the boundary between the street and the shop.

Brokula&Z 体验馆位于萨格勒布市中心，店面的设计始终围绕品牌的生态理念展开。店铺的橱窗内收，形成缓步台，供人休息，同时柔化了街道和店面的界限。

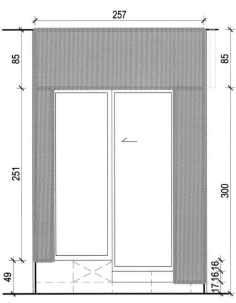

DC Store Shinsaibashi

心斋桥 DC 店

Design: Specialnormal Inc.
Location: Shinsaibashi, Osaka, Japan
Completion: 2013
Photography: Koichi Torimura

店面设计：Specialnormal 股份有限公司
店铺地点：日本，大阪，心斋桥
竣工时间：2013 年
图片摄影：鸟村弘一

To make the most of an original glazed façade, the designer intended to design something to draw passersby's attention. Not only a black panel signage attracts people's attention from a distance, but also the brand logo and the displays are both effectively revealed when the shop is lighted up in the evening. The image next to the brand logo can be changed seasonally in order to give different looks to the shop.

为了充分利用原有的玻璃店面设计，设计师计划打造一个能吸引路人注意力的新店面。其中黑色面板的使用有助于远距离吸引潜在消费者；夜幕降临时，店面照明设计使店面标志和商品展示更加直观。品牌标识旁的图像可以根据当季的营销重点进行改变，打造多变的店面外观。

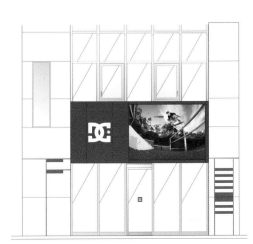

Donna Karan New York Store

唐娜·凯伦纽约

Design: Bonetti Kozerski Studio
Location: Cheongdam, Seoul, Korea
Completion: 2011
Photography: Lee Chul (Chul Studio)

店面设计：伯内蒂·科泽斯基工作室
店铺地点：韩国，首尔，清潭洞
竣工时间：2011 年
图片摄影：李哲（李哲工作室）

The store is located in the upscale Cheongdam shopping district of Seoul. The store occupies the first two floors of a six storey office building. The entire lower façade of the building was removed and a new façade was designed that incorporates large blocks of textured and chiseled granite. On the corner a large double height window was created by cutting away part of the second floor. This new glass façade is framed with large panels of black glass in sleek contrast to the textured stone. A new doorway into the store is located to the right of the double height show window. To the right of this entry is a double height, 10 foot wide LED video screen built into the façade. This screen will play a variety of media; from the latest Collection image video to a live link to the latest Donna Karan show during New York Fashion week. To the left of the display window, along the narrower alleyway by the side of the shop is a vertical bamboo garden that climbs up the solid granite of the store. This wall has punched openings through it which give glimpses into the store within.

这家店面位于首尔的清潭洞高档购物区。店面占据了六层中的一、二两层。工程对建筑低层的整个外墙都进行了重新设计和施工，新外墙设计使用了大块的花岗岩。这些花岗岩纹理清晰，轮廓分明。去掉二楼的一部分墙面后，设计师在店面的转角处打造出了两层高的大橱窗。新的玻璃外墙以黑色玻璃为框，与石材形成强烈的反差。店面新入口位于双层橱窗右侧。入口右侧是嵌入墙壁的两层楼高、10 英尺（约 3 米）宽的 LED 显示器。显示器会播放一系列影音内容，既有新系列的介绍视频，又有纽约时装周上唐娜·凯伦秀场的最新画面。橱窗左侧，紧挨店面旁狭窄通道的是一个垂直花园，植物沿着坚实的花岗岩向上攀爬。这面墙壁留有空隙，可以窥见店内的风光。

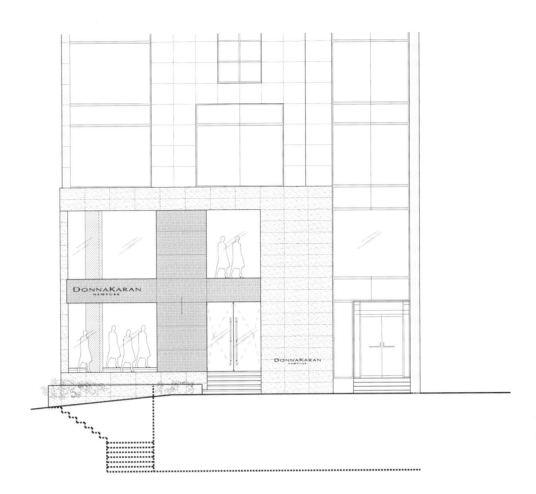

Family Center, The Store

家庭中心服装店

Design: Ali Alavi
Location: Mazandaran, Iran
Completion: 2013
Photography: Ali Alavi

店面设计：阿里·阿拉维
店铺地点：伊朗，马赞德兰省
竣工时间：2013 年
图片摄影：阿里·阿拉维

There are a few reasons why the designers enclosed the long (30 meters) front wall. One was that the existing building's elevation was really unpleasant, and designers had no permission to touch it. By enclosing the front façade, they gained 30 meters of wall space inside.

Second, according to the daily observational statistics study, majority of shoppers chose to go inside the store just because of curiosity, and to see what's happening behind this façade. Curious forms.

Third, the main idea comes from nomadic architecture of the region with similar materials and contexture. Wood ceiling plank and panels, sloped roof (gable/ shed/gambrel), are all used in many of the buildings in Northern Iran.

本案中，设计师选择为店面增加 30 米长的外墙有以下几个原因。

首先，建筑原有的外墙不甚美观，而相关规定禁止对其进行任何改动。在此基础上，这样的设计增加了 30 米的店面空间，不失为一个创造性的举动。

其次，根据相关统计研究，大部分消费者会仅仅因为好奇进入店铺，探寻墙壁另一侧的世界。

再次，本案的设计思想主要源自当地使用了相近材料和质地的游牧风格建筑。木质天花板及其他板材，斜顶都是伊朗北部常见的建筑元素。

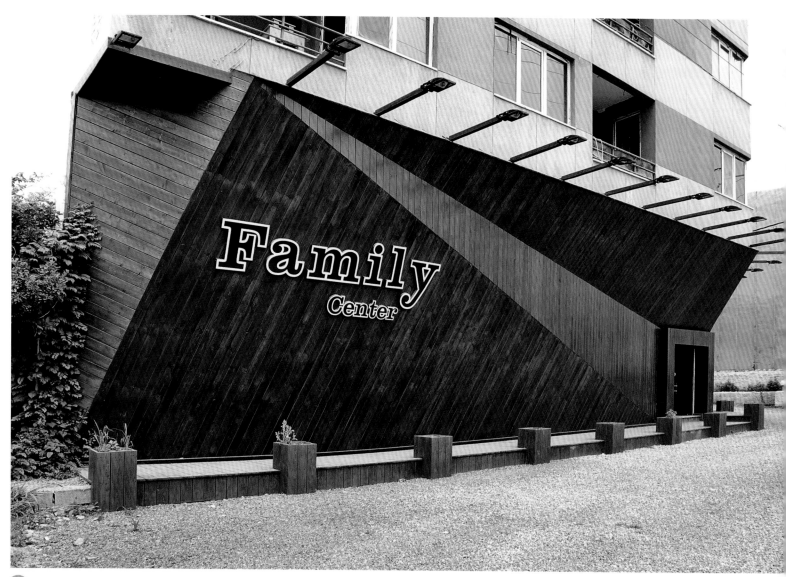

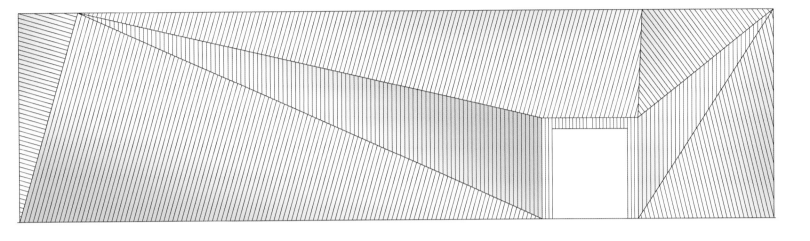

Ferrari Store

法拉利旗舰店

Design: Massimo Iosa Ghini / Iosa Ghini Associati
Location: Madrid, Spain
Completion: 2012
Photography: Ferrari and Nicola Schiaffino

店面设计：马西默·尤萨·基尼 / 马西默·尤萨·基尼设计工作室
店铺地点：西班牙，马德里
竣工时间：2012 年
图片摄影：法拉利品牌，尼古拉·斯基亚菲诺

The Ferrari Store in Madrid is of crucial importance to the Ferrari branded retail sector, as it will be the first branch to be reworked in accordance with the new store restyling concept. The clear protagonists of the new design are the Ferrari product and the brand's history.

Architectural input from the Iosa Ghini Studio focused on enhancing the windowed frontage, which is used to communicate the Store's racing spirit and tradition through the Fan space with the Formula 1 car and a window display dedicated to the Lifestyle world.

The highest quality finishes and materials, along with the utmost attention to detail, reflect Ferrari's soul, in a display which offers a balance between products and memorabilia which exudes the spirit of the Formula 1 'reds'.

The Ferrari Storefront doesn't just want to be a normal façade, it wants to be the design which reflects the Ferrari spirit and the brand's history, which conveys the dual soul of the racing world and a luxury universe.

马德里的法拉利旗舰店在法拉利品牌的零售部门中占有重要地位，因为它是第一个按照新店方案实施了改造的店面。法拉利品牌产品和品牌历史是新设计理念中的核心内容。

马西默·尤萨·基尼设计工作室进行的建筑改造关注临街店面的强化建设，利用风扇设计，一级方程式赛车和日常用品的橱窗展示传递品牌的赛车精神与传统。

商品展示在主流产品和纪念品之间达到平衡，选用高质量的表面材料与精致的细节处理更是呈现法拉利品牌的"红色"精髓。

法拉利旗舰店不止是一个店面设计，它是反映法拉利品牌精神和悠久历史的设计，具有赛车世界和奢华空间的双重灵魂。

Global Style Tokyo

Global Style 东京店

Design: PROCESS5 DESIGN/Ikuma Yoshizawa, Noriaki Takeda, Tatasuya Horii
Location: Tokyo, Japan
Completion: 2011
Photography: PROCESS5 DESIGN

店面设计：PROCESS5 DESIGN 设计公司，吉泽玖磨，武田纪昭，堀井龙哉
店铺地点：日本，东京
竣工时间：2011 年
图片摄影：PROCESS5 DESIGN 设计公司

The first Tokyo branch of Global Style, a made-to-order suit shop, has opened in Kanda, the textile quarter of eastern Tokyo. The exterior walls are a door-like façade produced from incombustible wood with the shop logo that express an entrance to a made-to-order suit shop designed by our company in the past. This is a design which was planned to have a part of the door penetrate into the interior of the building in order to attract customers who are walking past on the pavement in front to come into the shop. The colour of the wood of the exterior walls and the furniture inside is bright in order to sweep away the solemn image of traditional made-to-order suit shops.

Global Style 是一家定制西装店，在东京的第一家分店选址在东京城东部的纺织品中心——神田区。店面外墙使用类似门板的防火木材制成。店铺标识的主题是西装定制店的入口，也是由 PROCESS5 DESIGN 设计公司在之前的合作中完成的设计。在这个设计方案中，门的一部分穿透外墙，进入室内，以达到吸引顾客的目的。外墙木料和室内家具的颜色较浅，颠覆西装定做店略显严肃的传统形象。

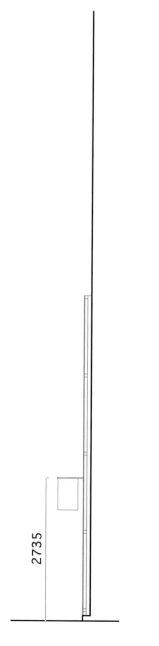
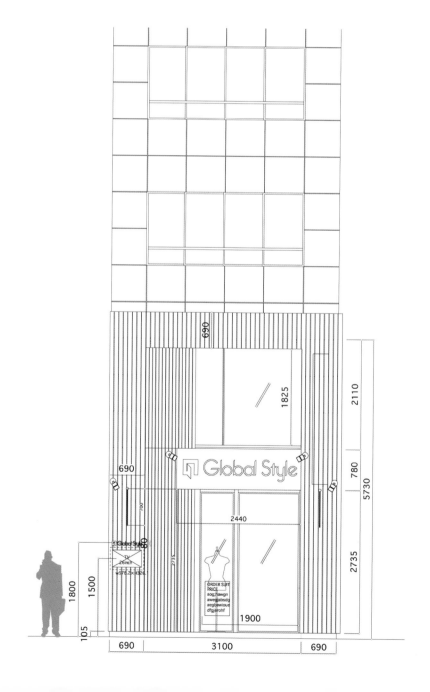

Sportalm Flagship Store Wien

维也纳斯博塔旗舰店

Design: BAAR-BAARENFELS ARCHITEKTEN
Location: Vienna, Austria
Completion: 2010
Photography: Petr Vokal, Michael Alschner

店面设计：巴尔·巴伦菲尔斯建筑师事务所
店铺地点：奥地利，维也纳
竣工时间：2010年
图片摄影：彼得·沃卡尔，迈克尔·阿尔施耐

Conversion of the new shop concept and complete redesign of the façade with glass and solid surface.
A complete redesign of the façade, which corresponds perfectly with the materials and the design of the interior, makes the overall picture of the new shop design perfect. The construction of the façade is a reticulated plane structure stiffened in a pixel-like fashion and plastic formed solid surfaced mould piece consisting of Corian and glass. The solution was an abstract, dynamic formulation of a snow landscape that would 'break' out of the uniform wall structure. This produced a form which would house supporting devices for clothing, positioned over recessed lights, as well as creating shelving surfaces.
The result was reached by accentuating a placidity of the wall surface, forming a fluid, curved plane. A homogeneous skin is created using white coloured Staron throughout and this colour is only broken where dark timber strips appear recessed within the shelving. The darkness of the timber forms a stark contrast to the white Staron material and serves in reinforcing light and shade of the wall curvature.

本项目是对店面新理念的实践，利用玻璃等材料对外墙进行了改造。
设计师对外墙进行了全面的重新设计，与室内装饰的材料及风格相呼应，构成完整统一的店面新形象。外墙原本采用的是网状平面结构，塑料表面有可耐力人造大理石和玻璃装饰。针对原有的店面条件，设计师创造了一个抽象的、具有动感的雪景，从单一的墙壁结构中突破而出。雪景的结构设计能够容纳服装支撑装置，提供置物空间。
改造工程通过增加墙体表面的平整度，形成流线型的曲面外墙。白色"星容"亚克力材料构成均匀的外墙，嵌入深色木材分散其间。木材的颜色与白色亚克力材料形成鲜明的对比，对墙面的立体感有一定的加强作用。

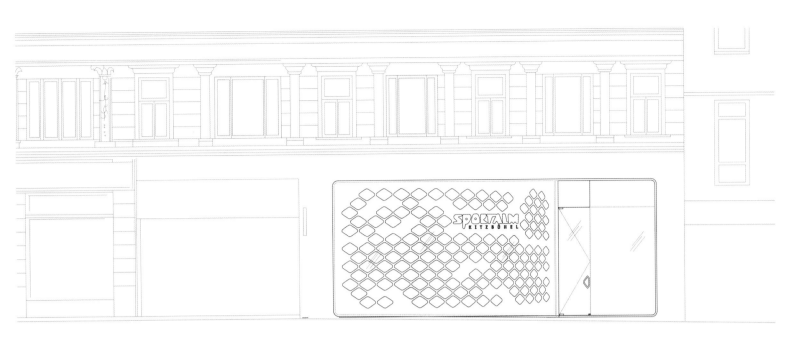

PART 8 Works-Fashion

Tienda Maria Cher

玛利亚·谢尔服装店

Design: Mathias Klotz arquitectos
Location: Buenos Aires, Argentina
Completion: 2013
Photography: Claudio Manzoni

店面设计：马赛厄斯·克洛兹建筑师事务所
店铺地点：阿根廷，布宜诺斯艾利斯
竣工时间：2013年
图片摄影：克劳迪奥·曼佐你

The project is simply a decision to generate the maximum exterior surface in an essentially interior site.
The Maria Cher store is in the Palermo Viejo district of Buenos Aires, a zone of urban renewal where commercial and residential convene in an environment of small to medium scale buildings built up side to side.
The façade is between two shared site walls and can be described as wide. The programme needed to make the most of the total 'constructability,' raising the land value to its maximum. To achieve all of this, along with the general haste of the project, designers created a store façade covered with wood to accentuate its natural character and conceal its irregularities.

本项目最大的挑战是要为一个有限的店面创造最大的外墙面积。
玛利亚·谢尔服装店位于布宜诺斯艾利斯的巴勒莫区。这里的中小型建筑排列紧密，由于城区改造工程，这里是商业和住宅用房的聚集地。
店面改造工程需要充分发掘场地的"可施工性"。为了实现这一目标，同时考虑到工期短任务重，设计师选择引入一个钢结构框架，对木质结构和服装展示起到支撑作用。楼上的几层结构都以金属波纹板覆盖。界墙以木材覆盖，体现亲近自然的特色，也能起到掩盖不美观建筑结构的作用。

ASICS Flagship Store

ASICS 服装旗舰店

Design: Verdego
Location: New York, USA
Completion: 2009
Photography: Augusta Quirk Photography

店面设计：Verdego 设计公司
店铺地点：美国，纽约
竣工时间：2009 年
图片摄影：奥古斯塔·夸克摄影

VERDEGO designed the first ASICS stand-alone store in the U.S. for its technologically advanced shoes and clothing line. The store is strategically located across from Bryant Park in New York City and is a vibrant center for the running community with the New York Marathon as one of the biggest races it serves.

ASICS, a Japan based company, wanted a sleek and bold façade design for the intimate 1100 square foot space. VERDEGO's response was a clean, sharp and minimalistic façade design that spoke to the companies' brand and aesthetic.

VERDEGO took its cue from the brand – reflecting the branding intense blue colour. The designers used the same deep penetrating blue of the logo on the sculptural wall that adorns the main wall of the store that is visible from the storefront.

VERDEGO 设计公司为 ASICS 品牌设计了在美国第一家独立店铺，并且负责宣传技术领先的鞋履和服装产品。店面选址在布莱恩特公园对面，是纽约城内跑步运动最活跃的区域之一。ASICS 是一个总部位于日本的服装品牌，该品牌要求设计一个时尚、大胆的店面设计。设计师依据品牌精神和风格量身定制了一个简洁、大胆、简约的店面设计方案。

VERDEGO 设计公司从品牌精髓——浓郁的蓝色——中汲取设计灵感，设计师在墙面使用了与品牌标识相同的蓝色，突出品牌特色。

Engelbert Strauss Workwear Store

恩格尔贝特·施特劳斯制服商店

Design: plajer & franz studio gbr
Location: Bergkirchen, Germany
Completion: 2012
Photography: die photodesigner.de

店面设计：Plajer & Franz 设计工作室
店铺地点：德国，贝格基兴
竣工时间：2012 年
图片摄影：die photodesigner.de 摄影

The working environment together with its emotional values forms the basis for the store design created by berlin based plajer & franz studio. The ingenous realisation evokes faith and tangibility for the consumers. at the same time it expresses specialist competence. 'craftsmanship' is real at the engelbert strauss workwear store and not simply a means of decoration as common in lifestyle- and fashion stores. it is a 'real' store for 'real' craftsmen who are looking for high-quality workwear and appreciate a surprising product display and shopping atmosphere. Emotional focus points are integrated into the store design with the intention to structure the space and create a customer journey.

工作环境和情感价值构成了 plajer & franz 设计公司柏林工作室这项店面设计的基础。独创性的设计有助于激发消费者的忠实度和品牌的切实感。同时，它还表达了专家般的独特魅力。与普通的时装店不同，顾客可以在这里切实感受到"工艺"的存在。这个"真实的"店铺为"真正的"工匠们提供高质量工装、充满惊喜的商品展示以及流连忘返的购物氛围。设计师也将情感的关注点融入了店面设计，尝试构建一个能够产生共鸣的购物空间。

Dikeni Menswear Boutique

迪柯尼男装精品店

Design: Stefano Tordiglione Design Ltd
Location: Yingkou, Liaoning Province, China
Completion: 2013
Photography: Stefano Tordiglione Design Ltd

店面设计：斯特凡诺设计有限公司
店铺地点：中国，辽宁省，营口
竣工时间：2013年
图片摄影：斯特凡诺设计有限公司

Befitting a setting more sophisticated than its home in Yingkou, China, Dikeni's flagship store marks a growing trend for luxury menswear brands in China and its pioneering design, courtesy of Stefano Tordiglione Design, is set to blaze a trail ahead of further developments. A striking 31-meter long and 8-meter high façade prepares customers for the massive 1,000 square-metre space inside the Dikeni store. Spanning two floors, the project was a challenging undertaking, but its realisation is the embodiment of tastefulness and refinement. The storefront mixes shiny and matte metal strips in a dynamic manner to make the façade stand out, while the verticality of these lines makes it appear taller and more imposing. Large windows are interspersed with impressive marble columns and black drops that hint at the luxury and exclusivity of the store and its products within.

营口的迪柯尼旗舰店是中国高端男装品牌发展的一个标志。斯特凡诺设计有限公司开创性的优质设计为品牌的发展开启了全新的方向。
旗舰店的店面外墙长31米，高8米，气势磅礴，称得上是1,000平方米的超大室内空间的预演。这间迪柯尼旗舰店跨越两层楼，设计充满挑战。最终效果代表了品位和精致，是对品牌精神的最好诠释。
外立面由闪亮和亚光的金属饰条组成，极具动感，垂直的线条使店铺看起来更加高大、雄伟。特大的橱窗点缀着独具特色的大理石柱和黑色吊饰，含蓄地表现商店及其产品的档次与奢华。

Hangzhou Flower Enjoy Store

Flower Enjoy 杭州武林路店

Design: Guan Design/ Jian Zhang
Location: Hangzhou, China
Completion: 2013
Photography: Jieyu Liu

店面设计：杭州观堂设计（张健）
店铺地点：中国，杭州
竣工时间：2013年
图片摄影：刘宇杰

With the philosophy of 'elegant woman', Flower Enjoy emphasizes on the concept of 'elegance'.
The shopfront features a black and white colour scheme. White is the main colour for walls, and black is employed for elements that can be seen from outside through the window, such as shelves, display table, lounges and cashier desk. The interaction between spaces helps reveal the essential concept of the brand.

"优雅知性女人"是Flower Enjoy品牌的核心，因此店面设计时，需要更多突出"雅"的氛围。
店面色系上以黑白为主旋律，墙体采用白色，橱窗内透出的道具类如货架、展示桌、休息沙发、收银台等选择黑色，搭配使用，整体简洁明快，突出Flower Enjoy品牌的淑女核心。

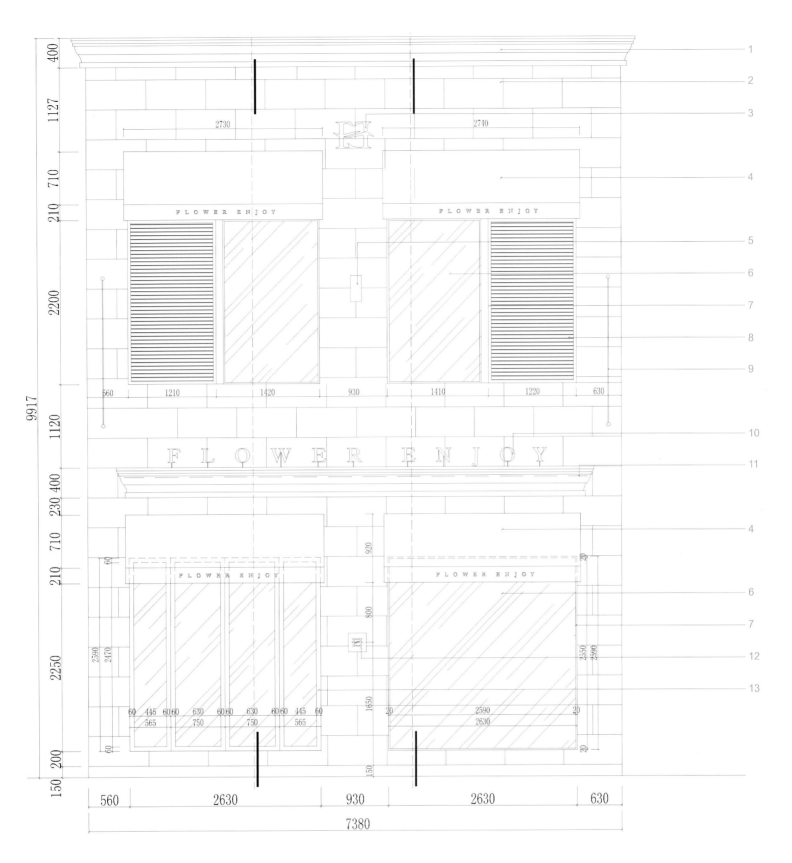

1. Customized stone features
2. 400×820mm travertine (sample)
3. Black backlit stainless steel logo
4. Black canopy (metal frame)
5. Wall lamp (sample)
6. 15mm tempered ultra clear glass
7. 20×60mm black window frame with electrostatic coating
8. Black aluminum blind with electrostatic coating, harbouring air-conditioning units
9. Black wrought iron bar
10. Black backlit stainless steel lettering
11. Hidden lighting devices
12. Black backlit stainless steel logo with electrostatic coating
13. 60×60mm black window frame with electrostatic coating

1．定做石材线条
2．400×820毫米 洞石（选样）
3．黑色背发光不锈钢浮雕 logo
4．黑色雨棚（金属框架）
5．壁灯（选样）
6．15毫米钢化超白夹胶玻璃
7．20×60毫米方管黑色喷塑窗框
8．铝百叶黑色喷塑，内藏空调外机
9．黑色铁艺挂杆
10．黑色不锈钢背发光字
11．内藏照明装置
12．黑色不锈钢喷塑背发光 logo 板
13．60×60毫米方管黑色喷塑窗框

Filles du Calvaire

受难修女街服装店

Design: Laurent Deroo Architecte
Location: Paris, France
Completion: 2012
Photography: C. Weiner _ LDA

店面设计：劳伦·德鲁建筑师事务所
店铺地点：法国，巴黎
竣工时间：2012年
图片摄影：C·维纳

For this project, Laurent Deroo Architecte office planned to settle the shop window in front of building masonry in the way to have a unique material outside: basaltina. The choice of this lava stone allowed to have a mineral texture close to concrete and to give at the same time, a noble status to the shop. Thereby, the front shop shows a strong contrast between opaque and transparent parts, between the cold colour of stone outside and the warm colour of wood inside. To increase this contrast, the entrance door has been treated as a monumental part that belongs to this opaque surface, made of the same stone. Here, glazing acts as holes opening views towards inside spaces. To avoid classical architectural language, the stone has been installed in a very simple design. The only formal treatment is a rounded cutting edge radius to soften the ensemble. Basaltina finish has a very thin orange peel effect that gives stone a pleasant smooth touch. It has also been installed on shop floor and furniture to have material continuity from outside to inside, associated to brass and monvingui wood. Air conditioning grids have been designed as thin and long horizontal slots, below glazing, to keep front shop unity.

设计师计划在店面外墙使用一种特殊的材料：巴莎天娜瓷砖。这种熔岩石料质感与混凝土相近，但能够为店铺增添高贵的特质。因此，店面不透明与透明的部分，室外岩石的冷色和室内木材的暖色形成了强烈的反差。为了加强这种反差，设计师将入口大门作为不透明表面的一部分处理，使用了相同的石材。玻璃部分引导消费者关注室内。为了避免使用传统的建筑语言，石材部分的图案设计十分简单。唯一的处理手段是圆形的边缘切割，起到柔化整体效果的作用。石材表面的橙色薄膜赋予了石材一种光滑的触感，地面和家具也使用了相同的材料，配合黄铜和木材获得室内外和谐统一的效果。空调网格采用的是水平方向低于玻璃结构的狭长设计，利于保持店面的整齐统一。

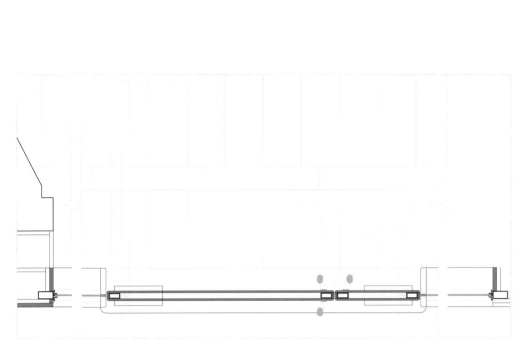

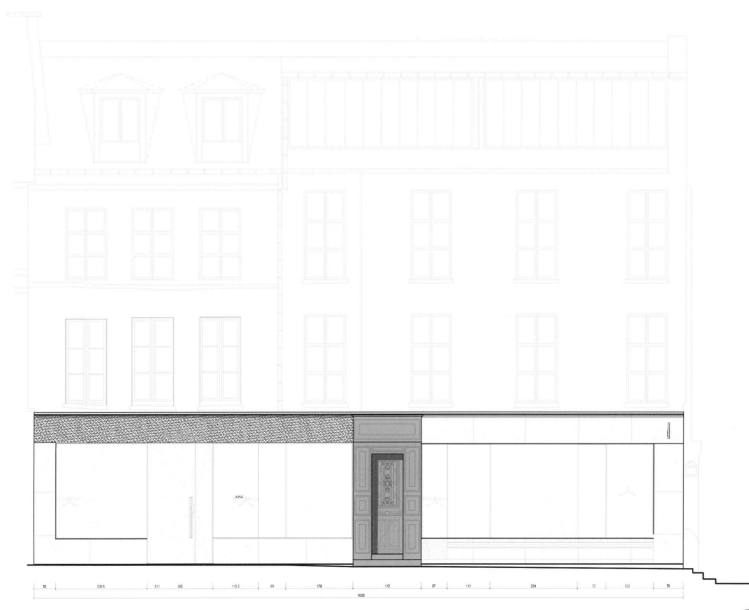

Glassons Flagship Store

格拉森斯旗舰店

Design: Studio Gascoigne, Naomi Rushmer, Mark Gascoigne, Theresa Ricacho and Wallace Ong
Location: New York, USA
Completion: 2010
Photography: Patrick Reynolds

店面设计：加斯科因工作室，内奥米·拉什莫，马克·加斯科因，特蕾莎·里喀秋，华莱士·翁
店铺地点：美国，纽约
竣工时间：2010年
图片摄影：帕特里克·雷诺兹

The Glassons flagship store in Auckland's main fashion precinct of Newmarket was inspired by the theme of an elegant mansion. A large single window was chosen to allow for large, theatrical displays with a solid backdrop that enables theatre-style backdrops to reinforce the windows theme.

The window is literally 'framed' with an ornate but hard-wearing picture frame finished in an aged faux-pewter. The entry Portico features wrought iron gates, decorative ceramic floor tiles and a ceiling dome, which all add to the sense of grandeur.

Customers love the playful touches that are incorporated in each window display and this has been very effective in drawing passers-by into the store's interior.

这家位于奥克兰商业区的格拉森斯旗舰店以优雅的豪宅为设计灵感。设计师选用大扇窗户进行宏大、夸张的橱窗展示，纯色背景使这种剧场般的橱窗主题得到加强。
窗口采用了华丽但耐磨的人造白蜡边框设计。铁艺大门、陶瓷装饰地面和穹顶天花板为入口门廊增添了富丽堂皇的感觉。橱窗展示中的趣味细节深受顾客的喜爱，也能够有效地吸引路人进入店内。

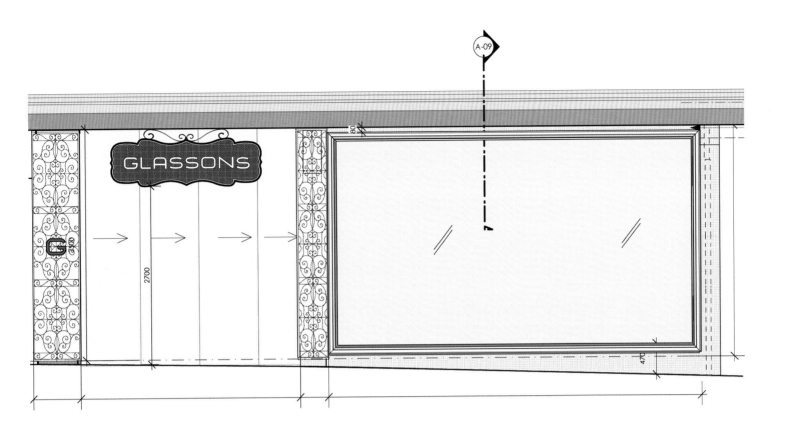

1. Fire indicator panel
2. Exrerior fram
3. Lobby wall
4. Window display
5. Steel panel

1. 消防指示板
2. 外框
3. 大堂墙壁
4. 橱窗展示
5. 钢护板

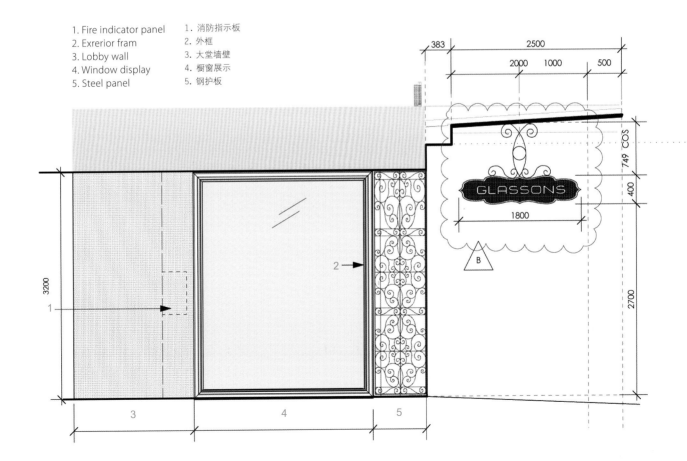

Ningbo GXG Jeans Flagship Store

gxg jeans 宁波城隍庙旗舰店

Design: Guan Design/ Jian Zhang
Location: Ningbo, China
Completion: 2011
Photography: Fei Wang

店面设计：杭州观堂设计（张健）
店铺地点：中国，宁波
竣工时间：2011年
图片摄影：王飞

GXG jeans is an energetic clothing brand for men, and its shopfront design emphasizes the brand philosophy of comfortable fashion, featuring an attitude for passion, innovation, and self-consciousness. The 'book' concept of the flagship store symbolizes lifestyle with stronger personality and better taste. From the façade, extending to the display window and interior, all are themed around books. One can find in here white models of books, books published in other countries, photos of books, and even book themed wallpaper, flooring, and ceiling. This helps generate a multiple effect for the space.

gxg jeans 品牌男装专为都市青年量身定做，强调年轻人多彩的生活姿态。店铺设计上，着重从服装风格出发，力求营造自然、清新、优雅的风格，追求激情、创新、自我的气质。将店铺的主题定义为"书"，年轻的一代，有活力，更有内涵，书，是对生活的追求，对品位的追求。因此，店铺从外立面开始，延伸至橱窗，再到室内，充满书的影子，形式纷呈多变——有白色的书模，有真实的外版书，有书的照片，更有书的墙纸、地面、天花，丰富多彩却不夺人眼球。

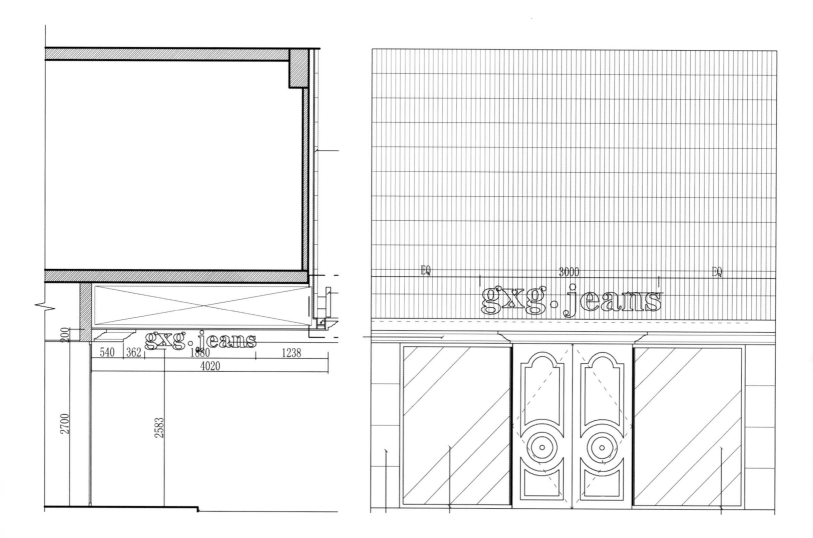

Volcanic Slab D2C Flagship Store in Hangzhou

"地壳运动"——D2C 杭州旗舰店

Design: 3GATTI/Francesco Gatti
Location: Hangzhou, China
Completion: 2012
Photography: Shu He

店面设计：3GATTI 设计公司 / 弗朗西斯科·加蒂
店铺地点：中国，杭州
竣工时间：2012 年
图片摄影：舒赫

D2C is a multi brand platform to buy the most interesting international fashion brands on the internet and now also on a physical space, a challenging physical space. Volcanic Slab is an existing building slab reshaped with volcanic energy to accommodate two floors of retail space. The concept is as simple as this but creates a very intricate and complex labyrinthic space that will challenge your shopping experience. This simple idea creates a very strong identity of the shop; even coming from the street is possible to see immediately the reshaped slab on the building façade. This squared skyline is not only a façade decoration but you will notice immediately from the glass window that it is actually the shape of the slab that will continue inside the store.

立足互联网的国际设计师品牌营销平台，D2C 如今有了线下模式——一个极具趣味的实体空间。"地壳运动"的概念十分简单，亦即在现存楼体中模拟地心热能所产生的地壳运动对楼板进行重新塑形，以此构造一个两层楼的购物空间。空间一方面复杂多变，形似迷宫，所提供的购物体验更为新奇、有趣，更具挑战性。另一方面，也有很强的识别度。行人走在街上便能对重塑的楼板一目了然。其高低曲折构成的线条不仅对外观构成装饰，也使顾客从店外透过玻璃墙便能明确看出店内的这种模仿地壳运动的形式。

Karl Lagerfeld Store, Berlin

卡尔·拉格菲尔德柏林旗舰店

Design: plajer & franz studio gbr
Location: Berlin, Germany
Completion: 2013
Photography: plajer & franz studio gbr

店面设计：plajer & franz 设计工作室
店铺地点：德国，柏林
竣工时间：2013 年
图片摄影：plajer & franz 设计工作室

Karl Lagerfeld opens the first German concept store in Berlin. With an overall size of 110 sqm the store is located in the trendy shopping area around Hackescher markt, in Berlin mitte. The 'mitte' quarter is vibrant, young, and upcoming, which fits the KARL LAGERFELD brand very well and communicates its idea - to be true accessible luxury.
The idea of the design is an encounter between French elegance and the unique glamour of the German capital. Huge Berlin style double-wing doors make a great appearance in the store, simultaneously giving the space a feeling of a classic apartment and separating the shopping area from the fitting rooms. Further, open ceilings and exposed brick walls slurried with a very thin white plaster break through the overall elegance of the store thus appearing very Berlin-like.

本项目是卡尔·拉格菲尔德品牌在德国开设的第一家旗舰店。店铺总面积110平方米，坐落在柏林中心区靠近哈克广场的时尚购物街上。这是一个充满活力、青春洋溢的前沿时尚区，非常贴合卡尔·拉格菲尔德的品牌精神，同时传达了"奢华而不失亲切感"的品牌讯息。
设计理念巧妙地结合了法式优雅和独特的德国风情。巨大的德国式双翼门占据了店面的主要位置，自然地呈现出经典的公寓设计感，并将购物区和试衣间隔开。进入店内，极细的白色石膏造型将开放的棚顶和裸露的砖墙联系在一起，柏林风在整体的优雅氛围中呼之欲出，令人印象深刻。

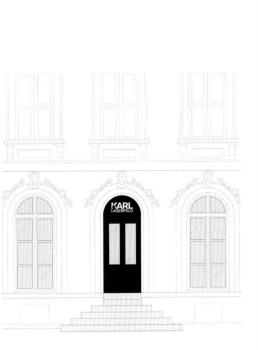

Jocomomola Store

伊都锦服装店

Design: churtichaga+quadra-salcedo architects
Location: Madrid, Spain
Completion: 2012
Photography: Elena Almagro

店面设计：churtichaga+quadra-salcedo 建筑师事务所
店铺地点：西班牙，马德里
竣工时间：2012 年
图片摄影：埃琳娜·阿尔马格罗

The shop occupies an old Hairdresser's shop, in this centric Madrid District.

The designer tried to give a certain 1950's atmosphere to this little ambience place, playing with vintage furniture, lamps and materials.

It's a little street but almost corners with a main commercial street, so the strategy on the façade was to make a shouting sign with the name of the shop. The designer painted the big letters in neon colour retro illuminated, and the designer reproduce in the rest of the front the traditional aspect of the town's old shops, avoiding showing any technical items as modern lamps or AC machines.

伊都锦服装店位于马德里中心区域，原本是一家理发店。设计师尝试使用古董家具、灯具和辅助材料在店内创造一种 20 世纪 50 年代的复古氛围。

店面位于一条不起眼的小街，邻近马德里的主要商业街，因此设计师认为外墙设计应主推店铺的名称和标识，吸引注意力。除了霓虹色字母的复古照明设计，店面的其他部分均采用传统的店铺设计风格，没有选择其他现代设备或照明装置。

Liu-Jo

Liu-Jo 服装旗舰店

Design: Arch. Christopher Goldman Ward, Christopher Ward Studio
Location: Reggio Emilia, Italy
Completion: 2011
Photography: Stefano Camellini

店面设计：克里斯多夫·高德曼·沃德，克里斯多夫·沃德工作室
店铺地点：意大利，雷焦艾米利亚
竣工时间：2011 年
图片摄影：斯特凡诺·卡莫里尼

The first Concept 'B' store is located in Reggio Emilia, a mid-size town in the northern part of Italy with a population of around 200,000. Reggio Emilia is also the hometown of Christopher Goldman Ward, the Italo-American architect in charge of Liu-Jo brand retail development.

Perfectly set in an 18th century building of the main street downtown, the boutique detail's accuracy can be noticed starting from the façade's air-conditiong system hidden by grills executed in 'Mondrian' style and subdivisions. Clamoring big metal mash cubes dress the shop window display, capturing the eye from the street, gently supporting last collection 'must haves', enflaming visitor's buying instinct.

Liu-Jo 服装的第一家"B"级概念店位于意大利的北部的中型城镇——雷焦艾米利亚，人口 20 万。也是设计师克里斯多夫·高德曼·沃德的家乡。这位意大利裔建筑师对 Liu-Jo 服装的概念店进行了全面的设计。

本案位于雷焦艾米利亚市中心商业街上的 18 世纪古老建筑内，隐藏在格栅后的空调系统等装饰细节体现了设计师的创意巧思。店面橱窗使用的是大型金属网块结构引人注目，传达最新产品的低调高质主题，激发顾客的消费热情。

Joules

焦耳斯服装

Design: Checkland Kindleysides
Location: London, UK
Completion: 2010
Photography: Keith Parry

店面设计：Checkland Kindleysides 设计公司
店铺地点：英国，伦敦
竣工时间：2010 年
图片摄影：基斯·帕里

Checkland Kindleysides' new store for Joules, situated in Covent Garden is the brand's first standalone store in London and the first of a new retail concept inspired by contemporary country living - living the good life.
Newly built store features the same exterior architecture with glazed arched storefronts reaching nearly 8m high. Joules makes the façade with a bronzed fascia bearing the Joules identity. Whilst the façade is flanked by two large windows displaying the latest from Joules' ranges and allowing clear views beyond into the store.

checklandkindleysides 设计公司受邀为焦耳斯服装品牌在伦敦的第一家独立店铺进行店面设计。这家概念店位于科文特花园，采用的是以当代乡村生活为灵感的的全新零售理念——享受美好的生活。
新店面设计延续了所在建筑的外墙元素，玻璃店面高度接近 8 米。青铜色横带上是品牌标识。店面外墙的两扇大窗展示最新商品，方便行人了解经营内容和店内情况。

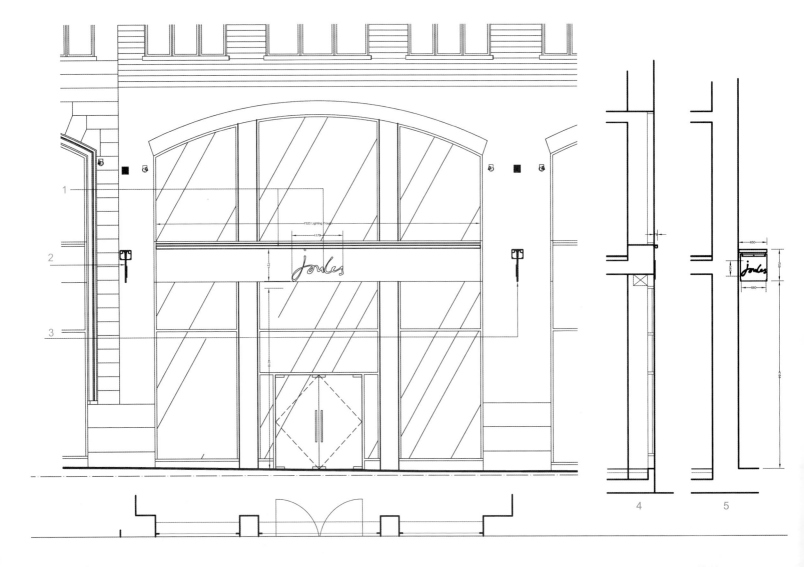

1. Main fascia sign
2. Existing projecting sign add Joules logo detail to inside face
3. Projecting sign
4. Section through main fascia
5. Section at Projecting sign

1. 横带标识
2. 原有指示牌以及品牌标识
3. 店面标识
4. 横带截面
5. 标识截面

Karl Lagerfeld Flagship Store, London

卡尔·拉格菲尔德伦敦旗舰店

Design: Plajer & Franz Studio
Location: London, UK
Completion: 2014
Photography: Karl Lagerfeld

店面设计：Plajer & Franz 设计工作室
店铺地点：英国，伦敦
竣工时间：2014 年
图片摄影：卡尔·拉格菲尔德时装

The latest addition to an international rollout of Karl Lagerfeld stores is located in the heart of one of Europe's most prestigious shopping areas - Regent Street - home to numerous well-known brand stores and fashion boutiques and a destination for 70 million visitors a year from all over the world.

Marked by simplicity, contrasts in surfaces as well as an interplay between black & white, the façade is created and implemented by Plajer & Franz Studio under the artistic direction of Karl Lagerfeld. The façade concept combines the brand's accessible luxury collections with cutting-edge digital technology as well as Lagerfeld's iconic design aesthetic thus blending edgy modern and classic elements. Materials and their surfaces play a vital role in the shop, often creating multidimensionality. Reflecting surfaces or other glossy surfaces contrasted with matted materials such as white laminate paneling on the walls allow playing with perspectives and thus the creation of subtle visual illusions – a creative wink and a detail becoming visible only on second sight.

卡尔·拉格菲尔德品牌旗舰店的最新成员坐落于欧洲最负盛名的购物区之一——伦敦摄政街。这里聚集了众多知名品牌旗舰店和时装精品店，每年接待来自世界各地的游客7000万。店面设计方案简洁大方，Plajer & Franz 设计工作室在卡尔·拉格菲尔德本人的指导下，通过表面材质以及黑白配色的反差对比，打造出一个现代与经典装饰元素混搭的成功案例。外墙设计结合了品牌的奢华风格和顶尖的数字技术，完美展现了卡尔·拉格菲尔德品牌标志性的美学设计。
建筑材料和表面质感在店面设计中起到了十分重要的作用，增加了空间的多维性。反光表面和白色墙壁板材的亚光表面构成对比，形成视觉多样性，让人驻足回味……

Kerry Center Concept Store

嘉里中心概念店

Design: Gruppo C14 srl
Location: Shanghai, China
Completion: 2013
Photography: Gruppo C14 srl

店面设计：Gruppo C14 设计公司
店铺地点：中国，上海
竣工时间：2013 年
图片摄影：Gruppo C14 设计公司

Shanghai's façade is designed in absolute creative freedom, it's a business metropolis par excellence, that continuously modifies its urban structure in accordance with the opening of malls and business centers; an oriental New York in a confused state of pro-capitalism, in which culture and traditions seem to disappear in the general noisy eagerness. In this sense it's a stretched skin, an impressive wave of cylinders, a pure expression of Peuterey's logo, as the first and immediate representation of the brand's values which determines its identity and recognizability, it becomes three-dimensional and superficial at the same time. The light draws the façade and follows its soft curves, constantly suggesting new interpretations, both visual and emotional. The façade is generated by a mesh of decorative bronze brass bars that describe a moving wave, accounting for 15 m of the façade while a videowall describes the logo's identity.

The development of the façade not only stems from formal reasons but is also the result of analysis of the dynamics of flows and routes with the aim being to create a casing that follows the flows themselves. The sinuous and dynamic movement of the façade cladding functions as an attractive element which draws one inside.

上海是一个不断变化的大都市，新开张的商场和商业中心不断地改变着它的城市面貌。上海概念中心的店面设计洋溢创意与自由的精神。某种意义上，店面外墙是一层延展的皮肤，一个令人印象深刻的波状图案，向世人展现品牌的价值观和独特个性。这是一个立体丰满的形象。店面造型会随着光线变化呈现不同的深度和曲线，为人们带来视觉和情感的新意。外墙由网状装饰黄铜棒组成，外形如同起伏的波浪，横跨15米的距离。大屏幕讲述品牌的故事。外墙设计不仅隆重美观，也是动态流向和路径调查研究的成果。这样的设计不仅功能性强，其店面外观能够吸引潜在顾客。

Levi Strauss, Berlin

柏林里维斯旗舰店

Design: Checkland Kindleysides
Location: Berlin, Germany
Completion: 2008
Photography: Checkland Kindleysides

店面设计：切克兰德·金德利赛斯设计咨询公司
店铺地点：德国，柏林
竣工时间：2008 年
图片摄影：切克兰德·金德利赛斯设计咨询公司

The three storey, 400m² store is located in the heart of Berlin, with a prominent position on Kurfürstendamm. With its striking façade of arched windows, spanning the full 3 storey height of the store, every floor is showcased at a single glance, this is further enhanced by the use of anti-reflective glass which provides crystal clear visibility into the store.

The store expresses the different personalities of the brand, creating a strong definitive area for each, sympathetically brought together to create the ultimate expression of Levi's®.

这个 400 平方米的三层建筑位于柏林市中心，位置得天独厚。店面外墙采用横跨三层楼高度的拱形窗户设计，令人印象深刻。使用防反射玻璃极大地提高了店面的通透性。

店面设计表达了品牌的多重个性，并将其成功整合，最终呈现出里维斯品牌的风格与特色。

Lucien Pellat-Finet Shinsaibashi

吕西安服装店

Design: Kengo Kuma & Associates
Location: Osaka, Japan
Completion: 2009
Photography: Daici Ano

店面设计：久万健吾合作公司
店铺地点：日本，大阪
竣工时间：2009年
图片摄影：阿野太一

In the meeting at Shinsaibashi, looking down the street of luxurious brand shops, Lucien asked for a soft and warm space, rather than icy, solid one. In response to his idea, designers propose a façade to realize the softness of Lucien Pellar-Finet cashmere in the architecture.
In seeking balance between the cost and the creation of various organic patterns, a 'vegetable wall' was born, which is made of structural plywood with two kinds of width and three types of aluminum connectors. In response to the 'vegetable wall' in the window, it use the aluminum to shape the same 'vegetable wall' to keep the consistency. It also used the purple steel platform to make the warm feeling.

吕西安服装店位于大阪心斋桥，邻近众多奢侈品牌店面。店主要求设计出一个柔和温馨的店面环境，拒绝冰冷僵硬的感觉。为了实现这一目标，设计师提出了一个开士米羊绒般柔软的店面设计方案。
设计中最特别的是由两种宽度的结构胶合板及三种铝连接器构成的"植被墙"，在成本开销与创造不同有机模式之间寻求平衡。为了与"植被墙"相配合，窗口处还设计了与其相同的铝制框架。紫色结构也增添了温馨的气氛。

Matalan

马塔兰服装店

Design: Checkland Kindleysides
Location: High Wycombe, UK
Completion: 2010
Photography: Matalan

店面设计：切克兰德·金德利赛斯设计咨询公司
店铺地点：英国，海威考姆勃
竣工时间：2010年
图片摄影：马塔兰服装

For this first site using the 'store of the future' concept for Matalan, the designers created impact from the outset, transforming the site, as the desigers enveloped the exterior in aluminium with contrasting red signage wraps. whilst the entrance area has been magnified with an over-sized red archway projecting from the building and the dramatic, double-height windows feature jumbo sized bill-boards and theatrical window displays.

本项目是马塔兰服装品牌第一个实践"未来商店"概念的店铺。通过多层级的设计，对场地的改造，设计师打造出由铝材覆盖的外墙，辅以红色标识形成鲜明对比。对入口区域进行了放大处理，突出的超大红色拱门以及夸张的两层楼高的窗户，充当特大的海报和夸张的橱窗展示。

Miss. Li

Miss.li 品牌女装

Design: Guan Design/Jian Zhang
Location: Hangzhou, China
Completion: 2013
Photography: Yujie Liu

店面设计：杭州观堂设计（张健）
店铺地点：中国，杭州
竣工时间：2013 年
图片摄影：刘宇杰

Miss.Li is a new women's clothing brand by Nawain. It is a story-telling brand of the founder herself. Miss Li had been courious and passionate about everthing related to fashion as a little girl. For her, the image of a perfect woman is gentle, charming, romantic, sweet, compassionate, optimistic and innovative in her career.

The shopfront design focuses on a young, passionate and positive attitude. Designers used windowpanes to decorate the façade. With a colour scheme of elegant ivory, it reveals the sweet, pleasant personality of a teenage girl. Among the various windowpanes, the big ones are glass-mounted so people can see the interior display from the street, and the smaller ones are decorated with sticker or grids to create a varied façade texture. This symbolises the diversified, curious brand philosophy of Miss.Li.

Miss.Li 是纳文品牌新推出的另一女装系列，顾名思义，Miss.Li 就是莉小姐，也就是关于其创始人自己的故事。莉小姐从小对一切关于服装的事物充满好奇和热情，在她脑海中，一个完美的女人形象，应该是柔和的、迷人的、浪漫的、甜美的、充满爱心的，对生活是积极乐观，对事业是敢于创新的。

店铺外立面设计重点在于渲染年轻的、热爱生活的、积极向上的氛围，设计师主要运用窗格来装饰立面。店铺外墙面选取优雅的乳白色为背景色，展现出少女甜美、柔和的性格特征。橱窗运用了不同形式的窗格，大面积的窗格选用玻璃镶嵌，便于人们在途径店面时直接看到室内的展示模特。其他小面积的窗格则选用贴纸或网格，形成多元且富有变化的外墙肌理。象征着 miss.li 品牌的多元化、对新事物的充满好奇与高接受度。

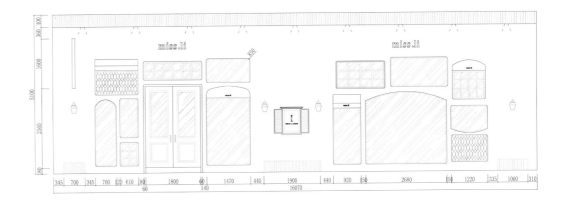

PART 8 Works-Fashion

Monki 3 Sea of Scallops

"扇贝之洋"服装店

Design: Electric Dreams
Location: London, UK
Completion: 2012
Photography: Electric Dreams

店面设计：电子梦想工作室
店铺地点：英国，伦敦
竣工时间：2012 年
图片摄影：电子梦想工作室

Launched in London, Stockholm, Oslo, Aarhus and Copenhagen, Monki head of store design Catharina Frankander speaks of the Sea of Scallops concept as built around the simple device 'more is more'. Although the design for the new store façades had budget limitations, designers decided to represent flexible systems, graphic guidelines, brand values, ambitious time plans, low watts per square meter, most importantly, a desire for novelty.

The façade is about story-telling theme. Designers used playful exaggeration, surprising shapes, and colourful graphics to create the surreal effect that is an important part of design concept. Every design detail in the new concept of shopfront emphasizes this. As usual the façade was all customized, for consumers are fascinated with the façade that is too colourful, too weird, too beautiful, too dark...

设计总监凯瑟琳娜·弗兰克安德用"多即是美"总结了"扇贝之洋"的设计理念。尽管店面改造项目存在预算限制，设计师决定使用灵活的工作系统，融合平面图样和品牌价值观，在短时间内完成对店面的翻新，以及对创新的追求。

外墙设计主题充满故事化的旋律，设计师选用活泼的夸张手法，奇异的形状以及色彩缤纷的图案营造超现实的效果。新设计中的每个细节都围绕并突出这一主题，这样的外墙定制方案使顾客对整个店面印象深刻。

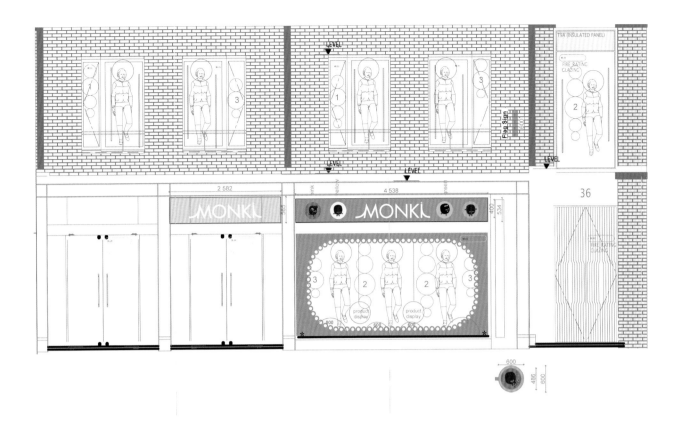

NICKIE in Lishui

丽水尼基儿童服装店

Design: Keiichiro SAKO, Shuhei Aoyama, ara Aghajani/ SAKO Architects
Location: Lishui, China
Completion: 2012
Photography: Ruijing Photo

店面设计：迫庆一郎事务所 / 迫庆一郎，青山修平，萨拉·阿噶嘉尼
店铺地点：中国，丽水
竣工时间：2012 年
图片摄影：瑞金摄影

In Alice's Adventures in Wonderland, Alice drinks the magical drink, which caused her to shrink, and then by eating the magical cake she grows to a tremendous size that her head hits the ceiling. In the story the changes in her size creates a kind of magical and fascinating atmosphere to the whole story. In this shop designers wanted to create a surprising space for the kids. Therefore designers took Alice Adventures in Wonderland as the reference and by exaggerating on the size of the elements they wanted surprise the kids. The reason that designers choose buttons to decorate the storefront, as the main element was that buttons have a direct relation with clothes and also learning how to button up clothes is a sign of growing up for a kid or in another words a kid can wear his/her own clothes without the help of an adult.

In order to keep the harmony in form, everything of the façade that was designed for this shop follows round and curvy shape. These curves are used in form of the arch in the entrance and the shop window.

在《爱丽丝梦游仙境》中，爱丽丝喝下了神奇的饮料，身体缩小，吃了魔力蛋糕之后身体增大，她的头撞到了天花板。主人公身体变小变大的情节为整个故事增添了神奇的迷幻色彩。在这个店面设计项目中，设计师主张为孩子们创造一个充满惊喜的奇妙空间。以《爱丽丝梦游仙境》为灵感来源，通过尺寸夸张的物品为小朋友制造趣味和惊喜。设计师使用纽扣作为店面的主要装饰元素，一方面是因为纽扣与服装有着直接的联系，另一方面这样的设计可以鼓励儿童学习自己穿衣系纽扣。

为了保持形式的一致性，店面外墙、入口和橱窗处使用了大量的曲线设计。

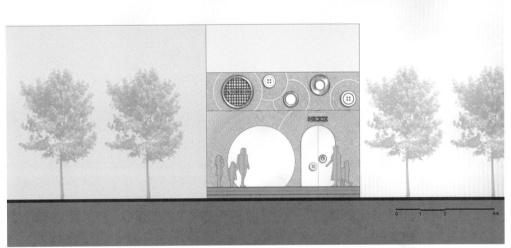

PART 8 Works-Fashion

Pili Carrera Boutique Flagship Store Madrid

皮利·卡莱拉马德里精品旗舰店

Design: MAM architecture, Pablo Menéndez
Location: Madrid, Spain
Photography: MAM architecture

店面设计：巴勃罗·梅内德斯，MAM建筑师事务所
店铺地点：西班牙，马德里
图片摄影：MAM建筑师事务所

The construction of this store in the Golden Mile of Madrid, the most exclusive area of Spain, aims to develop a new concept of retail space to project a new identity, a new image to the brand of children's clothing Pili Carrera. The strategy focuses on designing a children's clothing store for adults, discarding the much-touted option infantilizing selling space.

The project is presented as a façade with bright and warm materials that makes a neutral and elegant colourful container where exposed product. The showcase façade becomes an essential element. It does work not only as a product show window but also as a flared and protective filter between the outside and inside. This filter graduates and balances the delicate border between the warmth of the interior space and the public and cold character of the street. It is also relevant the relationship between the urban scale of the showcase and the scale of children's clothing. The flare generated by the sequence of interlocking slats of wood produces a transition between the large-scale of the showcase and small-scale of infant mannequins.

皮利·卡莱拉品牌的马德里旗舰店位于西班牙的黄金商圈内，设计师希望打造出一个具有全新理念的零售环境，为这个著名童装品牌赋予新的品牌形象。设计策略着重考虑在店内消费的成人的需求，摒弃了过于幼儿化的传统宣传手段。

店面外墙明快温馨，配色中性柔和。橱窗部分不仅具有展示商品的作用，还起到过渡店面内外空间的效果。这一结构对温馨的室内空间和冷淡的室外空间的微妙界限起到柔化、平衡的作用。纵横交错的木条完成了较大橱窗和较小的模特之间的转换过渡。

Size?

Size? 服装店

Design: Checkland Kindleysides
Location: Bristol, UK
Completion: 2009
Photography: Keith Parry

店面设计：切克兰德·金德利赛斯设计咨询公司
店铺地点：英国，布里斯托
竣工时间：2009年
图片摄影：基斯·帕里

Size?'s new store located on Bristol's Horsefair uses a concept which takes its inspiration from the locality.

The façade has an eclectic mix of materials and imagery to depict the heritage and modernity of Bristol. A focal point of the façade is an 1824 painting of the fair itself, which is emblazoned across the back wall of the store, whilst around the store mannequins wearing carnival masks reflect the popular attraction of 'freak shows' of this time. Designers used graffiti style graphics to reflect the present day and the quirky character of the Size? brand.

A timeless black and white photograph was applied to the storefront, confirming its Horsefair location. Window uses clear acrylic boxes for the presentation of T-shirts and trainers. At night roller shutters with tag graffiti are lowered behind the boxes in the window. Black and orange Signage is in playful space invader mosaic.

位于布里斯托市马市场的 Size? 服装店从店铺所在的独特位置汲取店面设计灵感。

在该项目的店面设计中，外墙设计混合使用了多种不同的材质和图案，充分描绘出布里斯托的文化遗产与现代文化的交融。店面设计中的一项重要内容是墙上一幅1824年创作的描绘圣詹姆斯集市的作品，店内带着嘉年华面具的人形模特与之呼应。设计师使用涂鸦风格的图案表现店铺的现代精神和独特个性。

店面的经典黑白照片形象带领人们回顾当年的历史。橱窗利用亚克力框架进行T恤和运动鞋的展示。黑色和橙色的店铺配色简洁醒目。

Quiksilver Lifestyle Store

极速骑板服装店

Design: Marcia Arteta Aspinwall, Fausto Castañeda Castagnino
Location: Peru
Completion: 2012
Photography: Zander Aspinwall

店面设计：杭马西娅·阿尔特塔·阿斯平沃尔，
　　　　　福斯托·卡斯塔涅塔
店铺地点：秘鲁
竣工时间：2012 年
图片摄影：桑德尔·阿斯平沃尔

In this project, environmental geography and climate play a crucial role in the design. The designers chose materials and textures accompanying the geographic location as well as the landscapes.

The old building's walls are coated with white and it counts with huge rimless glass doors and windows in order to make people feel cool and refreshed, which is also in accordance with the concept of the brand.

The cross linked wood skin takes possession of an ancient architecture which is timeless and inert. These cross horizontal and vertical at each side crossing again at the corner, in order to bring it to life with a sense of appropriation and identity.

This Quiksilver store was created to blend in with its surf town vibe and to make the brand provide a space for surfers to make it their own while communicating a lifestyle and an experience.

Fashion is like architecture. One dresses the men, the others the city.

本项目中，地理环境和气候对店面设计起到了决定性的影响。设计师特别根据当地的气候特点进行相应的材料和材质选择。设计师在原有建筑的墙壁上使用了白色涂料，配合宽大的无框玻璃门窗，给人以清爽的感受。这与极速骑板的品牌精神相辅相成。

木质外墙造型交叉相连，充分利用历史建筑的经典设计造型。木料在每个墙面横竖交叉，于拐角处再次交叉，为店面注入品牌独有的个性与活力。

极速骑板服装店的店面方案与这个冲浪小镇的环境氛围和谐相融，品牌为冲浪者提供了打造个性装备的空间，同时传递了一种生态态度与乐活体验。

时装就像建筑，装饰着人们，也装饰了城市的景观。

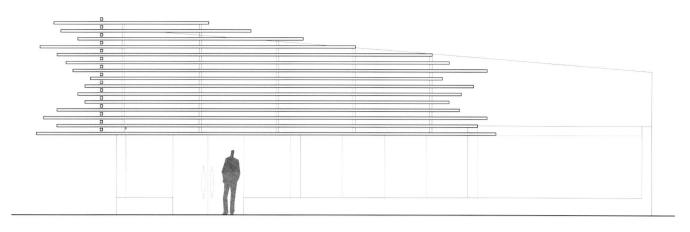
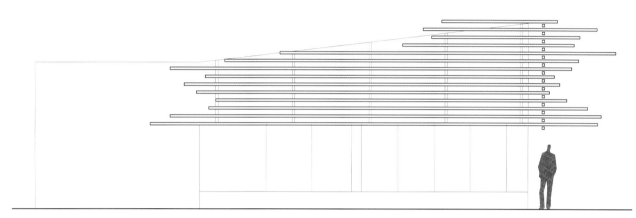

XG Store at Renmin Road, Wenzhou

XG（雪歌）温州人民路店

Design: Guan Interior Design / Jian Zhang
Location: Wenzhou, China
Completion: 2012
Photography: Wang Fei

店面设计：杭州观堂设计 / 张健
店铺地点：中国，温州
竣工时间：2012 年
图片摄影：王飞

Established in 1997, XG women's wear is committed to producing clothes of quality, elegance and exquisite detail. The brand image design project in 2012 fully respects the philosophy of simple elegance and reflects it in the shopfront across the country.

As the arch concept was proposed, architectural and decorative elements throughout the interior were decorated with curves and arches. The visual repetition forms uniformity and indicates the feminine and perceptual characteristics of XG women's wear.

Cement, white paint walls and wood were chosen as the dominant material, as a statement of simplicity and originality. It echoes the concept of the brand and adds a welcoming touch to it.

成立于 1997 年的雪歌品牌女装，一直致力于追求品质、优雅，强调细节设计。在 2012 年新一代店铺形象设计中，沿袭了雪歌一贯以来的质朴与素雅，并增添了贵族气质，形成别具一格的复古 + 现代的时尚混搭风格。

店铺设计中，提炼出"圆拱"的元素，从建筑顶部弧度处理，到货架的圆形弯拱，到装饰道具、试衣镜、背板的欧式圆弧线条等，由"圆拱"贯穿，用同一种元素将店铺形成视觉上的统一，同时，圆弧又能体现雪歌女装的柔美与知性。

材质选择上，主要采用水泥、白墙与木色搭配，其宗旨是追求材质的本色，摒弃一切花哨与浮夸的装饰，从而体现雪歌品牌一贯的坚持与追求——质朴简洁又不失温馨。

ZARA Kumamoto

ZARA 熊本店

Design: Studio ZARA / Key Operation Inc. / Architects
Location: Kumamoto city, Japan
Completion: 2010
Photography: Itou Prophoto Corporation

店面设计：ZARA 工作室 / KEY OPERATION 股份有限公司 / 建筑师事务所
店铺地点：日本，熊本市
竣工时间：2010 年
图片摄影：伊图摄影公司

This new building housing the Zara store boasts a main front that looks out onto a pedestrian street lined with traditional shops.
The front is a clever composition of horizontal and vertical elements that define the spaces for the windows, logos, etc. They project either inwards or outwards depending on their use, and features transparent glass, metal panels and sheeting for a friendly interpretation of the front. The skillful combination of vertical sheets reads like a bar code. The front was created exclusively for this project, but has opened up a host of possibilities for future designs.

ZARA 在熊本市的新店位于步行商业街内，沿街有很多传统店铺。
店面设计巧妙地利用横向和竖向结构界定了橱窗、品牌标识等元素。这些玻璃、金属面板构成的结构根据其所在的位置和功能，向内或向外倾斜。竖直方向的板材看起来像条形码一样，让人印象深刻。这个方案开创了店面设计的众多可能性，不失为一个经典案例。

The House of Kipling

吉普林之家

Design: UXUS
Location: London, UK
Completion: 2012
Photography: Dim Balsem

店面设计：UXUS 公司
店铺地点：英国，伦敦
竣工时间：2012 年
图片摄影：丁·鲍瑟姆

Kipling commissioned UXUS to create a global retail experience that embodies Kipling's playful personality and inspires its loyal customer following. The solution was to create a new home for the Kipling brand. Applying domestic charm to retail, 'The House of Kipling' takes a fresh perspective on everyday things.

Creating an instantly recognizable brand hallmark, the store façade features a signature brand pattern inspired by the iconic Kipling monkey. Uniform rows of colourful monkeys 'march' over Flemish style tiling, expressing the playful spirit and the heritage of the brand. Open windows invite clients to peek into and through the store, enticing them to step inside and discover the world of Kipling products.

吉普林公司委托 UXUS 公司为其进行全球店面的规划设计，体现品牌活泼的精神，吸引更多顾客加入使用吉普林的行列。"吉普林之家"的设计理念将家居用品的魅力借鉴到零售上，为日常用品赋予了新的内容。
店面外墙使用吉普林猴子的标志性品牌图案，彩色猴子图案整齐划一地排列在佛兰德风格的瓷砖上，体现活泼的品牌精神和品牌的传统文化。开阔的橱窗吸引顾客关注店内的商品展示，走进店里，探索吉普林的各种产品。

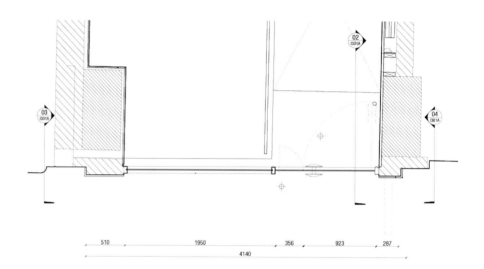

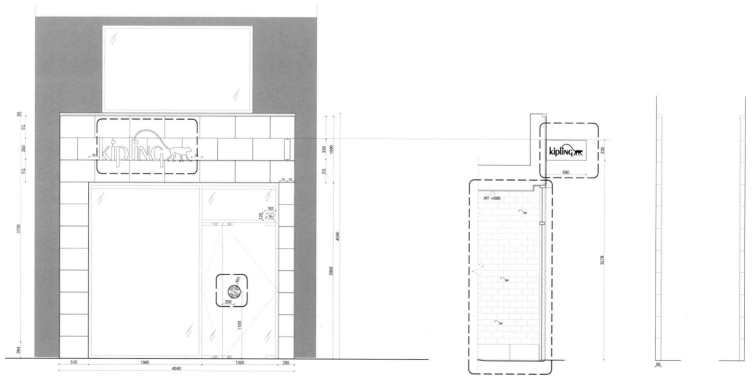

Goodwill Walla Walla

Goodwill Walla Walla 综合购物中心

Design: Tom Kundig, Les Eerkes
Location: Washington, USA
Completion: 2012
Photography: Kevin Scott

店面设计：汤姆·昆汀，莱斯·厄科斯
店铺地点：美国，华盛顿
竣工时间：2012 年
图片摄影：凯文·斯科特

The designers restored the building's natural beauty by introducing innovative daylighting techniques and stripping all excess materials from the building—revealing original brick walls, vaulted ceilings and wooden trusses. Natural light floods the interior by way of the new custom glass curtain wall located on the façade.

Sustainable strategies were also used throughout the design of this building, and served as cost-efficient solutions. The renovated façade is a structural fiberglass window system created from engineered wood and insulated glazing units set in sealant—creating excellent thermal performance.

设计师将多余的装饰去除，露出最初的砖墙、拱形天花板和木质桁架，并利用先进的采光技术，找回店面的自然美感。在店面外墙安装了玻璃幕墙后，店内获得了充足的自然光照。设计还充分融入了环保的理念和策略，成就高能效的设计方案。翻新后的店面变成了以实木和绝缘玻璃单元构成的玻璃纤维玻璃墙，具有极佳的保温性能。

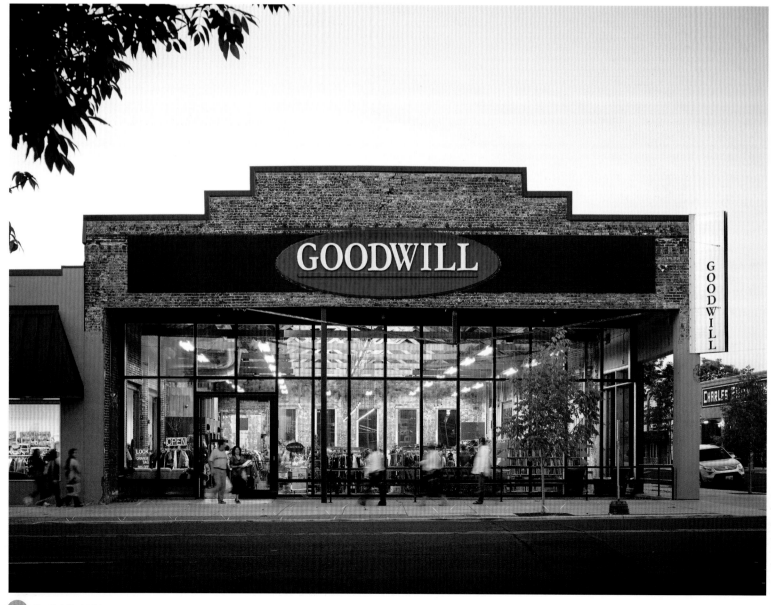

Donassy Open Atelier

唐纳西开放工作室

Design: Vanja Ilić Architecture
Location: Zagreb, Croatia
Completion: 2012
Photography: Vanja Ilić Architecture

店面设计：VanjaIlić Architecture 建筑设计公司
店铺地点：克罗地亚，萨格勒布
竣工时间：2012 年
图片摄影：VanjaIlić Architecture 建筑设计公司

The Donassy Open Atelier project created a temporary, flexible showroom whose purpose is to exhibit the work of the fashion designer Branka Donassy and other visiting artists. The existing storefront is in a historicist building, in a zone between Zagreb's Upper and Lower Town.

A black surface, made of rubber fabric, frames the entrance door, covering damage on the historicist façade, and marks the position of the specific content in the section facing the street. The interior space advances deep into the building like a tunnel, and is 'clothed' with a membrane made of stretch fabric, fixed to the wall with bolts in a dotted pattern.

唐纳西开放工作室的项目打造了一个现代、灵活的展示空间，主要容纳并展出时装设计师布兰卡·唐纳西以及其他艺术家的设计作品。场地位于萨格勒布上城区与下城区之间的一个古建筑内。

橡胶表面围绕入口大门，形成黑色的框架，掩盖古旧的外墙上留下的伤疤，也标记了店内对应位置的具体内容。室内空间向前探出，就像一个隧道，上方由固定在墙上的弹力织物覆盖。

New Look

New Look 女装品牌

Design: Checkland Kindleysides
Location: London, UK
Completion: 2012
Photography: Checkland Kindleysides

店面设计：切克兰德·金德利赛斯设计咨询公司
店铺地点：英国，伦敦
竣工时间：2012 年
图片摄影：切克兰德·金德利赛斯设计咨询公司

As the second largest women's fashion retailer in the UK and one of the most recognised High Street fashion brands, New Look strives to deliver fashion excitement, newness and value. The design brief from New Look was simple - create a store experience, which expresses the brand personality, which is flexible enough to adapt across small to large formats and which drives sales and advocacy.

At the Marble Arch store in London the corner staircase creates impact from the exterior and draws consumers into the store. The change starts from the exterior of the store; with a new, simpler, legible logo - reclaiming the best name in fashion retailing - and the added drama and impact visible in the window displays. These simple branding principles continue throughout the store, where the conversation is focused more on the core customer, women in their early 30s, with snappy headlines or lighthearted service prompts communicating the essence of the brand.

作为英国第二大女装零售品牌和最受认可的高街品牌之一，New Look 品牌传递时尚的兴奋、新奇和价值之感。这项店面设计的目标也十分明确：打造一个适应性强，刺激销售，且能充分表达品牌个性的店面模式。

New Look 品牌位于伦敦大理石拱门的店内，拐角处的楼梯构成一处独特的景观，吸引顾客进入店内探索。简洁、清晰的品牌标识彰显 New Look 品牌在时装零售业中的地位，橱窗展示增添了戏剧效果，引人注目。这些简单的品牌理念贯彻整个店面设计，针对三十出头的女性这一核心客户群体，以时髦的产品和热心的服务传达品牌的精髓。

Soloio Valencia

Soloio 巴伦西亚精品店

Design: Oficina Mutante
Location: Valencia, Spain
Completion: 2013
Photography: Oficina Mutante

店面设计：突变体建筑设计公司
店铺地点：西班牙，巴伦西亚
竣工时间：2013 年
图片摄影：突变体建筑设计公司

The interior illumination highlights the colour scheme, generating an overall image from the outside. The façade is minimalized and painted black thus creating a frame that, together with the opened doors, emphasizes the shopfronts. The doors, which remain open while the shop is closed, help frame the shopfronts, and at the same time protect the products displayed.

室内照明设计突出店面配色特点，突出店面设计核心。黑色店面外墙时尚简约，形成一个视觉上约束，构成店面的整体框架。玻璃展示外墙能在非营业时间保持店面的通透性，同时对店内展示的商品起保护作用。

Shoesme Tilburg

蒂尔堡 shoesme 鞋店

Design: Teun Fleskens
Location: Tilburg, The Netherlands
Completion: 2010
Photography: Michiel Maessen

店面设计：Teun Fleskens 设计公司
店铺地点：荷兰，蒂尔堡
竣工时间：2010 年
图片摄影：米歇尔·麦森

Shoesme is a high quality shoe brand for children, which handles the entire supply chain by itself, from production till distribution. The shopfront keeps the design philosophy of the dice on the outside. The designers used six dices of same size and different colours on the pedestrian road outside the shop, stacked irregularly as a symbol of the store.

荷兰 shoesme 是一个高质量的童鞋品牌，独立拥有从生产到经销的完整产业链。充满趣味的彩色骰子是店面的主要设计理念。设计师将 6 个大小相同、颜色不同的骰子安置在店外的人行道上，吸引消费者同时呼应店铺的主题。

Seraphita

塞拉菲达鞋店

Design: Stone Designs
Location: Madrid, Spain
Completion: 2013
Photography: Stone dsgns

店面设计：Stone 设计公司
店铺地点：西班牙，马德里
竣工时间：2013 年
图片摄影：Stone dsgns 设计公司

This project is the result of a deep study of the customer's buying experience and its relation with all the surrounding elements. In this way, the design tried to create different atmospheres that are perfect to show a wide range of shoes and bags.

The designers created an open wooden box, made in natural oak, which helps create a comfortable and sophisticated place, avoiding intrusive interventions and always considering the overall unity for a harmonious interpretation of the project.

本项目以一项有关消费者购买体验与消费环境关联性的深入研究为基础，旨在打造一个适合展示鞋子和提包的完美环境。设计师使用天然橡木制作了一个开放的木盒式结构，打造优雅舒适的环境，同时避免尖锐的干扰元素，将整体性与和谐性放在店面设计的首位。

Schutz Oscar Freire

奥斯卡弗莱雷街舒兹店

Design: Bel Lobo, Patricia Batista, Alice Tepedino, Ana Luiza Neri, Ayla Carvalhaes, Clarisse Palmeira, Fernanda Mota
Location: São Paulo, Brazil
Completion: 2012
Photography: Marcos Bravo

店面设计：贝尔·洛沃，帕特里夏·巴蒂斯塔，艾里斯·特贝蒂诺，安娜·路易莎·内里，艾拉·卡瓦海斯，克拉丽丝·帕尔梅拉，费尔南达·莫塔
店铺地点：巴西，圣保罗
竣工时间：2012年
图片摄影：马科斯·布拉沃摄影

This store façade was designed to evoke Schutz's strong identity at Oscar Freire Street, one of Brazil's most sophisticated shopping areas. To reflect brand's values, such as innovation and contemporaneity, Corian, usually used on interiors surfaces and furniture, was used as exterior cladding. White Corian plates were used in different depths, so that when back lighted they reveal different levels of translucency and opacity. The effect is an interesting play of light and shade, more subtle at daytime and more dramatic at night.

该项目店面的外墙设计意欲在巴西最繁华的购物街——奥斯卡弗莱雷街上为舒兹品牌打造出一个令顾客印象深刻的品牌形象。设计师选择在外墙上使用了通常用于室内装饰的可耐丽人造大理石材料，充分表达出舒兹品牌创新和现代的品牌精神。外墙上每块不同尺寸的白色可耐丽板材会在背光照明的设计下呈现出不同的透明度。白天看上去低调优雅，夜晚则动感十足。

Vass Shoes

瓦什鞋店

Design: T2.a Architects / Bence Turányi
Location: Budapest, Hungary
Completion: 2012
Photography: Zsolt Batár

店面设计：T2.a 建筑师事务所，本斯·土兰伊
店铺地点：匈牙利，布达佩斯
竣工时间：2012 年
图片摄影：若尔特·巴塔尔

The new shop is located in the heart of Budapest's city center, in the pedestrian zone Haris Köz. The timeless architectural and interior design of fine details and quality materials reflects on Hungarian craft traditions, the state-of-the art appearance of the façade refers to Budapest's classic shops from the beginning of the 20th century.

这家新鞋店位于布达佩斯市中心的哈里斯科斯步行街。经典的建筑风格和精致的室内设计代表了匈牙利的传统工艺技术，外墙的现代外观设计让人联想起20世纪布达佩斯街头的经典店面设计。

Camper Store, Granada

格拉纳达 Camper 鞋店

Design: A-cero Joaquin Torres & Rafael Llamazares architects
Location: Granada, Spain
Completion: 2012
Photography: Juan Sánchez

店面设计：A-cero 杰奎因·托雷斯和拉斐尔·利亚马萨雷斯建筑师事务所
店铺地点：西班牙，格拉纳达
竣工时间：2012 年
图片摄影：胡安·桑切斯

The location of the store is really good being placed in the most famous and commercial street of Granada. The store, with 48m², used to be a clothing shop partially abandoned. Two colours: white and red, typical of the A-cero interior design projects and Camper.
The front is made of composite panel with red aluminum and the showcase has been made of glass and red vinyl following the style of the store.

店铺位置优越，位于格拉纳达最著名的商业区。这个占地 48 平方米的店面曾经是一家服装店。

店面采用了 A-cero 建筑师事务所经典的红白双色设计方案，店面使用复合面板和红色铝材，展示柜用玻璃和乙烯树脂打造，构成和谐的店面风格。

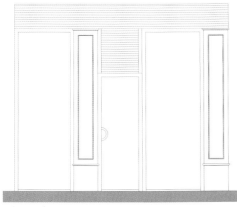

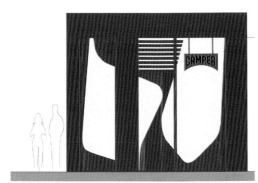

Alter Concept Store

凹凸前沿设计上海概念店

Design: 3GATTI
Location: Shanghai, China
Completion: 2010
Photography: Shen Qiang

店面设计：3GATTI 设计公司
店铺地点：中国，上海
竣工时间：2010 年
图片摄影：沈强

Alter is a project for an alternative fashion store. The design was fast and spontaneous, as usual Francesco Gatti designed like a child, without inhibitions. From the transparent storefront there is a natural gesture to develop a stair surface and at the same time exhibit the products in a multidimensional way.

The philosophy of Alter, as the word say is to be and inspire an alternative world, so as a designer Francesco Gatti imagined an alternative storefront like the ones in the drawings of Escher, where gravity and the rules of the normal world doesn't exist anymore, where there is no 'up' or 'down', no 'left' or 'right', and where everything is possible.

这是为一家时装店进行的店面设计，设计过程流畅而愉快。设计师弗朗西斯科·加蒂继续了他一贯的孩童般的创作风格，在透明的墙壁里创造了一个有力的连续空间，一方面将独特的阶梯平面持续延伸，另一方面以多维的手法进行商品展示。凹凸前沿设计的哲学理念，是创造并启迪一个非主流的世界。对于设计师弗朗西斯科·加蒂来说，一个非主流的建筑空间就好比荷兰画家埃舍尔的作品，是处于失重状态的。自然世界里的规则不复存在，这里没有上下左右之分，一切都成为可能。

FREITAG Store Tokyo

弗赖塔格东京店

Design: Torafu Architects
Location: Tokyo, Japan
Completion: 2011
Photography: Sebastian Mayer / Torafu Architects

店面设计：Torafu 建筑设计公司
店铺地点：日本，东京
竣工时间：2011 年
图片摄影：塞巴斯蒂安·梅耶 / Torafu 建筑设计公司

FREITAG, a Swiss brand of bags and accessories made out of recycled truck tarpaulins, bicycle inner tubes and seat belts of disused cars, opened its first Asian flagship store in Tokyo. The interior and exterior were planned with FREITAG's 'recontextualize' principle in mind, where new roles and different functions replace the purposes originally intended.

Located on a corner site in Ginza and occupying the first two floors of what used to be an old shoe shop in a 50 year old building, the project was envisioned to be 'more than a store'. The designers had aimed for the store to be an extension of the street, of which the brand is characteristically originated from. In order to allow the liveliness of the surrounding streets to naturally enter into the store interior, the large glass panels making up the storefront can be fully slid opened. Breaking down the conventional shop typology even further, an existing window opening on the first floor was made to resemble and function like a 'drive-thru' window, taking communication with the street to a more personal level. Occupying the main wall of each floor inside the store, the individual products are housed in the 'V30 FREITAG Skid' patented shelving units which are used in all FREITAG stores. The products naturally emerge as the focus of the simple interior.

瑞士品牌弗赖塔格主营由回收卡车篷布、自行车内胎和废弃的汽车安全带制成的箱包和装饰品。在东京的这个项目是弗赖塔格品牌在亚洲开设的第一家旗舰店。

店铺位于银座一条街道拐角处有 50 年历史的大楼之内，占据一楼和二楼。这里曾经是一家鞋店，而本项目的设计目标是打造一个"不只是店面"的工程。设计师希望将店面设计成为街道延伸出的一部分，作为品牌起源的缩影。为了将周围的繁华和活力以自然的方式引入店内，设计师选择使用大扇的玻璃窗作为店面外墙，而且玻璃窗可以自由滑动，完全打开。在此基础上，设计师还尝试对传统的店面设计进行进一步的颠覆，一楼原有的一扇窗户开口就像快餐店的"来得速"窗口，进一步加强了店铺与外部环境的沟通和互动。

PART 8 Works-Fashion

Alberto Piazza

阿尔贝托广场鞋店

Design: Vincent Choi, Ruth Tjitra
Location: Broadbeach Queensland, Australia
Completion: 2011
Photography: Mark Duffus Photography

店面设计：文森特·蔡，露丝·基特拉
店铺地点：澳大利亚，昆士兰宽阔海滩
竣工时间：2011 年
图片摄影：马克·达弗斯摄影工作室

Alberto Piazza shoes represent the reference point in Brisbane for 'Made in Italy'. The brief was to create a paradigm for a new shopping experience that acknowledged and converged the rich culture of Italy and distinctiveness of Australia.

A simplistic, refined approach has been adopted for Alberto Piazza. To minimise visual distractions from key items, joinery and supports have been concealed. A floating effect for the internal display units is used to produce a seamless look, uncompromised in structural integrity. The curtain behind the displays obscures any sightlines. Antique look frames are positioned around selected units to accentuate and define. Evening and occasion footwear are exhibited in these units. This gallery atmosphere envelopes customers and transports them to Italy. It is important to remind customers that the products are made in Italy. This statement is what separates this brand from many others.

阿尔贝托广场鞋店代表的是布里斯班的"意大利制造"。项目要求设计师将意大利悠久的文化底蕴和澳大利亚独特的人文风情整合在一起表现出来，创造出一个经典的新型购物环境。

阿尔贝托广场鞋店采用的是简约、精致的店面设计风格。为了最大程度地减少对中心物品的视觉干扰，设计师选择将精细木工和支撑结构隐藏起来。内部展示单元呈现的飘浮效果构成了一种浑然天成之感，完美地实现了空间一体性。晚宴鞋和休闲鞋等需要突出的特定单元处利用做旧框架进行强调。这种画廊般的艺术气息会让置身其中的顾客有亲临意大利的体验。告知消费者产品为意大利制造这一信息是很有必要的，因为这是阿尔贝托广场鞋店最根本的特色之一。

Di Marco Shoeshop

迪马可鞋店

Design: Manuel García Estudio
Location: Alicante, Spain
Completion: 2011
Photography: ClaroOscuro Fotografía

店面设计：曼纽尔·加西亚工作室
店铺地点：西班牙，阿里坎特
竣工时间：2011 年
图片摄影：克拉罗·奥斯库罗摄影

This commercial space corresponds to a second opening that this brand has carried out after the redesign of its corporate image. The project pursues to achieve three goals: to create shops which help to enhance the displayed product; to use a kind of materials which give personality to the interior space; and finally, to achieve product differentiation with regard to competition.

The original façade had an unflattering irregular shape which was not functional and that is why it was decided to modify the entire structure. The new entrance door has been moved to the left side achieving, in this way, a totally lineal window display. Above the entrance, an overhanging roof has been made respecting the original shape of the place. This creates a hallway which welcomes the client and where a back lit image of one of the main brand they offer has been placed.

On the other hand, from the right side of the façade the new wide and continuous window display emerges and whose composition lets the footwear be the real protagonist of the window display. This effect is multiplied due to the grey mirror covering the lateral party walls.

商业空间的设计主题呼应重新设计后的品牌形象。这一工程有三个预期目标：创造一个能够加强商品展示的店面；使用一种能够为室内空间增添个性的建筑材料；最后，实现产品的差异化竞争。

店面原本是一个不规则形状，实用性不强，急需对整个建筑结构进行改造。店面新入口改建在建筑的左侧，使橱窗展示直观而清晰。依据入口上方的形状添加了外伸屋顶，构成玄关结构、方便顾客的同时，也为店面主打产品提供了更精致的展示空间。

此外，新的宽阔橱窗从店面外墙的右侧延伸开来，使店内的鞋类产品成为展示的中心内容。横向界墙上的灰色镜面更加强了这种效果。

Nino Shoes Store

尼诺鞋店

Design: Dear Design
Location: Barcelona, Spain
Completion: 2013
Photography: LaFotográfica

店面设计：Dear 设计
店铺地点：西班牙，巴塞罗那
竣工时间：2013 年
图片摄影：LaFotográfica 摄影公司

The shop window has a double reading, as the furniture exhibition can be seen both from outside and inside. The design objective is to multiply the effectiveness of exposure meters. The shop window, enlightened by abundant natural lighting, participates as a display throughout the store, facing the street. Different finishes have been chosen to create a set of contrasts between white and black, and matte and glossy finishes.

本项目中的店面橱窗设计使人们在店面内外都可以看到商品展示架，主要的设计目的是加强橱窗的展示效果。自然光照下，橱窗框架整合了店内的展示内容，呈现紧凑充实的店面空间。使用不同表面构成黑白配色、光面与亚光材质之间的反差和对比。

Optic Shop Laskaris

拉斯卡瑞斯眼镜店

Design: dARCHstudio, Elina Drossou Architects
Location: Kifissia, Athens, Greece
Photography: Stathis Mamalakis

店面设计：伊琳娜·杜罗索，尼科斯·卡卡彻拉斯
店铺地点：希腊，雅典，凯菲西斯区
图片摄影：斯塔西斯·马摩拉西斯

Transparency, semi-transparency, opaque and reflection are the main characteristics incorporated in the façade design of the Laskaris Optics shop. The inspiration is drawn from the nature of the eye. Outside the shop passersby can be all eyes to the exhibition of the store's commodity and also have a clear view to what is happening inside. Along the windows customers can cast an eye over the spectacles, featured on the glass-shelves of the suspended large white Corian cylinders. The two façades become noticeable by the contrast of their materials. The opacity and the dark grain of wood contradict the transparency and the icy-hued colours of plexiglas and corian.

透明、半透明、不透明和反光材质的组合与碰撞是拉斯卡瑞斯眼镜店外墙设计中的主要特点。项目的设计灵感来自眼睛的生理结构。店面外的行人可以清楚地看到店内的情况。吊起的白色可耐力圆柱结构与玻璃展示架结合，向人们充分展示店内的商品。两面外墙所使用的木材和树脂玻璃的质感对比令人印象深刻：木料的不透明性和深色纹理与树脂玻璃的晶莹剔透反差强烈而奇妙。

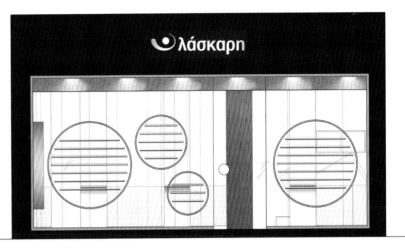

Kirk Originals Flagship Store, London

Kirk Originals 伦敦旗舰店

Design: Campaign
Location: London, UK
Completion: 2010
Photography: Hufton + Crow

店面设计：Campaign 设计公司
店铺地点：英国，伦敦
竣工时间：2010 年
图片摄影：霍夫顿与克劳摄影

Located in Conduit Street, in the heart of London's west end, the new store showcases the KirkOriginals collection of stylish glasses and sunglasses in its entirety along with selected ranges byindependent eyewear brands exclusive to Kirk Originals in the UK.

Taking inspiration from the brand's latest Kinetic collections, the flagship store design features displays of winking eyes in various guises. A series of larger than life lenticular printed eyes are suspended in the front window, simultaneously winking and catching customers' eyes as they approach and enter the store. A sense of interaction continues inside, as a wall display of human-like 'winkies' runs the length of the store, providing a ready-made audience to enhance the browsing and trying on experience. The 'winkies', 187 white powder-coated sculptural heads, each wear a unique frame and can be tilted and re-positioned to create clusters of onlooking craning heads.

A restricted palette of monochromatic colours and modest materials including blue-grey painted walls and a dark grey floor keeps the spotlight firmly on the 'winkies' adorned with frames as if displaying works of art.

本项目位于伦敦西区最繁华的康迪街，店内经营KirkOriginals品牌最新的时尚眼镜饰品以及独立眼镜品牌与之独家合作推出的仅向英国发售的眼镜产品。

该旗舰店的设计灵感来自于该品牌最新的Kinetic系列产品，使用了各式各样的眨眼图案。正面橱窗悬挂了一组夸张的眼睛一样的透镜，这样有趣的视觉效果有效地吸引顾客进店探索。这种互动的氛围在店内得到延续，187个白色头部雕塑分别佩戴不同的眼镜框架，突出品牌个性的同时起到展示商品的作用。

店面使用单一的配色和低调的建筑元素，如蓝灰色墙壁和深灰色地板等，使得人们的注意力始终关注"眼睛"的主题和相关产品。

PART 8 Works-Fashion

Prismóptica

棱镜眼镜店

Design: SPRS Arquitectura
Location: Oporto, Portugal
Completion: 2013
Photography: Rui Moreira Santos, Espinho

店面设计：SPRS 建筑师事务所
店铺地点：葡萄牙，波尔图
竣工时间：2013 年
图片摄影：鲁伊·莫雷拉

This store is located in Oporto, near to 'Museu de Serralves', is a small shop with forty square meters. The owner, an investor on the optics commerce, already had two other shops in two different cities, but with complete different concept.

The intention in this project was to create a space very similar to a jewel, and picking the idea of the name: 'Prismóptica', which means Prism+Optics, create a relation between the name and the proper shop. Every element on this shop was treated exactly like a piece of jewelry, and every detail was thought till the exhaustion.

The show window is composed for the object that supports and exposes the sun and optic glasses, and was inspired in the sense of a futuristic nature, like if the sun and optic glasses were fluting above the light on the steel tubes.

本店位于波尔图，是一家 40 平方米的小型店铺。店主从事眼镜业多年，已经在其他不同城市拥有两个设计理念完全不同眼镜店。

本项目的设计目标是创造一个珠宝一样的精致空间，体现店铺名字的寓意：棱镜＋光学的结合体，将店名与店铺本身切实结合在一起。设计师对店面的每个元素、每个细节都投入了大量的心血。

橱窗装饰元素传递太阳镜和光学眼镜的设计精髓，超越基本需求，充当日常生活的必需品。橱窗设计充分展示太阳镜和光学眼镜产品，具有很强的未来感。

Ray-Ban Store

雷朋眼镜店

Design: Arquitectura y Diseño
Location: Buenos Aires, Argentina
Completion: 2012
Photography: Julio Masri, Alejandro Hazan

店面设计：构架与设计工作室
店铺地点：阿根廷，布宜诺斯艾利斯
竣工时间：2012 年
图片摄影：胡里奥·马斯里，亚历杭德罗·哈赞

The premise for designing the store was to replicate Ray-Ban brand image both inside and outside by using the corporate colours, logos and graphic design. Arquitectura y Diseño designed a modern project with simplicity of lines and elements. From the outside, the store is highlighted because of Ray-Ban's signature red colour. Designers chose high quality paint for the wall coating. Under the big shop windows, stainless steel was used and the Ray-Ban lettering was debossed on the entrance floor, which shows the roughness and chic of the Ray-Ban brand. In the night the logo is shining in pink and white light, which is illuminated for calling public attention.

这项店面设计延续了雷朋一贯的品牌形象，无论室内室外都使用了标准配色、品牌标识和平面设计，打造一个由简洁线条和元素构成的现代店面设计。外墙使用高质量涂料，雷朋品牌的经典红色使店面尤为亮眼。橱窗下采用不锈钢结构，入口处地面上刻入了雷朋的品牌名称，彰显粗犷别致的品牌精神。店铺招牌上的标识会在夜间散发粉色和白色的光芒，吸引公众的注意。

Magic Store

奇幻隐形眼镜概念店

Design: Torafu Architects
Location: Omotesando, Tokyo, Japan
Completion: 2011
Photography: Daici Ano

店面设计：Torafu 建筑设计公司
店铺地点：日本，东京，表参道
竣工时间：2011 年
图片摄影：阿野太一

Torafu Architects designed the interior of a concept store that opened in Omotesando, Tokyo featuring Menicon's new line of contact lenses; Magic. Using the comprehensive contact lens manufacturer's unique Flat Packaging Technology—said to be the world's thinnest (1mm)—Magic represents a major departure from conventional contact lens packaging.

From the transparent storefront, the customer can see a ribbed display wall where the thin and compact packages can be shelved on any part of it. The wall, which wraps the interior of the store smoothly, has extrusions and depressions matching the size of these packages—like a white canvass on which to display products freely. With the transparent storefront, the store is made to look like one big showcase room that faces the street. The designers aimed to create a showcase-like space that gives off a feeling of novelty every time one looks at the ever-expanding colour variation of packages on display.

本项目是为隐形眼镜品牌美尼康在东京表参道的的概念店进行的店面设计，主推其最新的隐形眼镜产品：奇幻。这一系列产品使用的扁平封装技术，据称是全球目前最薄的（1 毫米）隐形眼镜包装技术，奇幻代表了区别于传统隐形眼镜包装的重大突破。

穿过透明的店面，顾客可以看到一个层次感很强的展示空间。墙壁的起伏设计作为展示空间，就像一块洁白的画布，充满无限可能。玻璃外墙使店面看起来像一个开放的巨大展示间。设计师希望这样的设计能够给人带来持久的新鲜感，留下深刻的印象。

Permy-mi Jang Won

Permy-mi Jang Won 美发沙龙

Design: Kim Rang, Kim Jae-jin / M4
Location: Gyeonggi-do, South Korea
Completion: 2011
Photography: Lee Pyo-joon

店面设计：M4 设计公司 / 金朗，金在英
店铺地点：韩国，京畿道
竣工时间：2011 年
图片摄影：李杓俊

'Permy' a compound word of 'perm' and 'mommy', is a cute girl with a perm that can be found in several areas both on the outside, emphasizing the identity of the space. The overall colour plan consists of only two colours-white and sky blue - to present a pure and pleasant feel.

This character is not only used as a simple graphic element of this space but also has the flow of a great story. The story is the processes of achieving Permy's dream to be a hair designer, to launch her own hair shop, to create new hair style and to win the prize at the competition for hair designers.

Filled with small but delicate details, Permy-mi Jang Won will be remembered thanks to its unique elements including the character and various accessories.

"Permy" 是一个正在上卷发棒的可爱女孩，她的形象出现在美发店内外的多个区域，深刻的强化了店铺的品牌形象。该美发店选择了白色和天蓝色作为主色调，为顾客呈现了纯净愉悦的感觉。

这个人物形象不仅作为简单的平面元素在店内使用，也构成了一个完整的故事：女孩实现了她成为发型师的梦想，开设了自己的美发店，创造出精美的新发型，赢得无数奖项。

人物形象的塑造和多种装饰的使用令这家美发店充满精致细节，让人印象深刻。

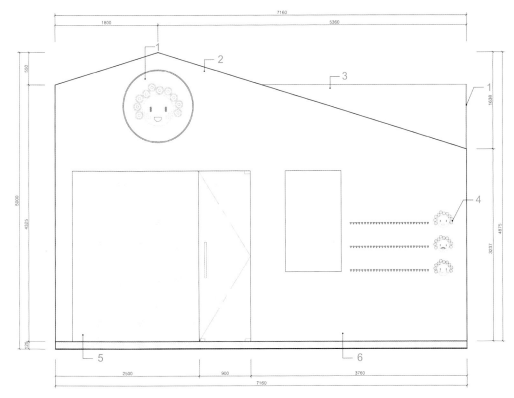

Elevation a
1. Sign
2. APP. with colour LAC'Q fin.
3. APP. Retromint LAC'Q paint fin.
4. APP. sheet fin.
5. APP. THK 12mm clear tempered glass fin.
6. APP. white colour LAC'Q paint fin.

Elevation b
1. APP. wastebasket install
   Propuction light
2. Mirror
   APP. retromint colour paint fin.
   APP. THK 10mm glass fin.
   APP. white colour paint fin.
   APP. retromint colour paint fin.
3. APP. white colour paint fin.

Elevation c
1. APP. wastebasket install
   Propuction light
2. APP. retromint colour paint fin.
3. APP. retromint colour paint fin.
4. APP. sheet fin.
   APP. white colour paint fin.
5. APP. tile fin.
   Floor level construct
6. APP. retromint colour paint fin.
7. APP. white colour paint fin.
8. APP. retromint colour paint fin.

Elevation a 立面 a

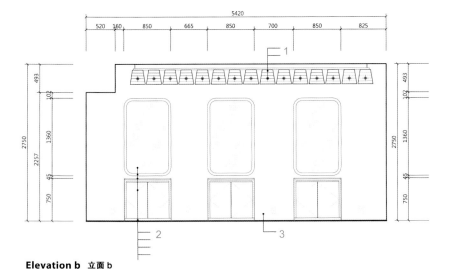

Elevation b 立面 b

立面 a
1. 图标
2. 彩色涂料表面
3. 复古薄荷绿涂料表面
4. 板材表面
5. 12 毫米厚钢化玻璃表面
6. 白色涂料表面

立面 b
1. 废物篮装置
   照明
2. 镜面
   复古薄荷绿涂料表面
   10 毫米厚钢化玻璃表面
   白色涂料表面
   复古薄荷绿涂料表面
3. 白色涂料表面

立面 c
1. 废物篮装置
   照明
2. 复古薄荷绿涂料表面
3. 复古薄荷绿涂料表面
4. 板材表面
   白色涂料表面
5. 瓷砖表面
   地面高度结构
6. 复古薄荷绿涂料表面
7. 白色涂料表面
8. 复古薄荷绿涂料表面

Elevation c 立面 c

Essential Hair

奥卡斯发型设计

Design: KC design studio
Location: Taipei, Taiwan
Completion: 2011
Photography: Ivan Chung

店面设计：康乾设计工作室
店铺地点：台湾，台北
竣工时间：2011 年
图片摄影：钟汶权

The layout of traditional hair salon usually has three spaces, reception, staff working and service. In this project, the designers decided to break the rule and reorganize the space to improve interaction with customers.

The design concept was to create a vivid and natural environment with industrial style at work by utilizing a number of suspended partitions in the middle of wall to perform an open circle to make customers have their privacy while being served, and add some green on the white wall to make the natural speaks.

The major material used was wood, followed by I shaped steel plates supporting the partitions. Moreover, the designers are expecting to impress customers at the front door with the big glass and stainless wall to provide a natural and cozy hair salon space.

传统的美发沙龙的布局通常分为三大部分：接待处、工作空间和服务空间。在这个项目中，设计师试图将这种传统的平面配置方式重新解构，打破原有的界限，创造出人与人之间交会的场所。

在空间概念上，利用数个悬浮在空间中的折板墙围塑出不同的平面使用分区；借由折板墙所折曲出的角度引导主要服务动线，成了人在空间之中流动的重要指引。同时，在折板的设计上结合了发艺造型时必备的镜台以及旧杉木板，其目的是将生硬的工作平台自然化，空间中塑造出有自然气息的舒适工作环境。

在副材料设定上，设计师采用了粗犷中带有细腻的 I 字型的薄钢板作为支撑折板墙的主结构，增加整体视觉上的穿透性。裸露的白砖墙和绿色植物点缀，增添空间活泼又有个性的气氛。最后入口大厅，以大面落地玻璃镜面与毛丝面不锈钢作为主要印象，期待消费者能感受到流行时尚感及自然舒适的环境。

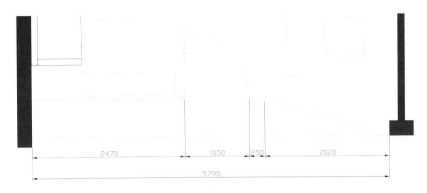

Hershesons Flagship Salon

赫施森美发旗舰店

Design: gpstudio
Location: London, UK
Completion: 2013
Photography: John Adrian
Awarded: Shortlisted for the Design Week Awards (UK)

店面设计：gpstudio 设计工作室
店铺地点：英国，伦敦
竣工时间：2013 年
图片摄影：约翰·阿德里恩
所获奖项：入围英国设计周大奖

A favourite of celebrities, models and the press, Hershesons tasked gpstudio with creating a luxurious customer experience in a space that reflected Daniel and Luke Hersheson's high-end fashion credentials.

This spans across two floors, and creates a central spine more akin to an installation in an art gallery or fashion catwalk, rather than a hair salon. The salon has been completely transformed, with a brand new look for the shopfront, offices and treatment rooms. A high quality, yet simple palette of marble, brass, wood and polished plaster is used throughout, creating a modern and very luxurious feel.

赫施森美发店深受明星、模特和新闻界人士的喜爱，gpstudio 设计工作室受邀为其进行店面改造，打造奢华的消费体验。设计应反映品牌创始人丹尼尔·赫施森和卢克·赫施森的高端时尚品位。

设计方案横跨两层楼，中央骨架更像是艺术馆和时装 T 台的装置。整个沙龙获得了彻底的改造，店面、办公区和剪发区都焕然一新。使用高质量，但配色简单的大理石、黄铜、木材和光面石膏使沙龙现代之中透露出奢华的感觉。

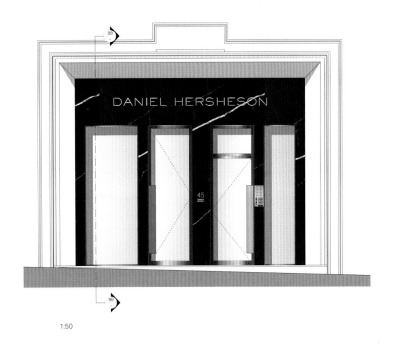
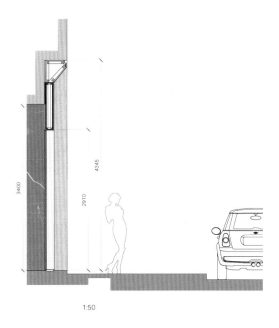
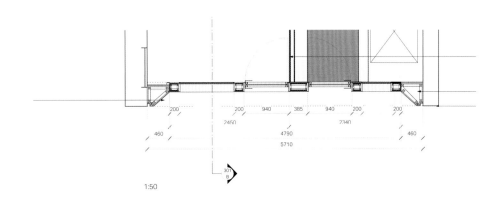

PART 8 Works-Beauty

How Fun Hair Salon

"好玩"美发沙龙

Design: J.C. Architecture
Location: Taiwan, China
Completion: 2013
Photography: Kyle You

店面设计：柏成设计有限公司
店铺地点：中国，台湾
竣工时间：2013年
图片摄影：游宏祥

The façade is a distinct interface between the bustling lane and the inside salon. The tunnel almost pipes right through the lane and offers that entire journey inside from the outside lane with glass separating the private and public space, the façade is a great, crisp, clean interface between the two spaces. The tunnel creates a distinct space that starts a journey for the customer. The tunnel helps transform the customer's direction into an intimate pathway across the salon floor allowing.

The black coated façade and white sign, wooden tunnel and door, concrete stair, the grass on the top of canopy give people a kind of natural and fresh.

本项目中的店面外墙构成了喧闹的街道和沙龙内部之间的屏障和交界面。通道深入店铺，利用玻璃起到分隔公共空间与私人空间的作用。外墙在店铺内外空间内构成了完整、通透的交界面。通道的设计构成了一个独特的空间，为顾客提供印象深刻的消费体验。

店铺的黑色外墙与白色标识，木质通道和门，混凝土楼梯以及屋顶花园等元素营造出天然清新的氛围。

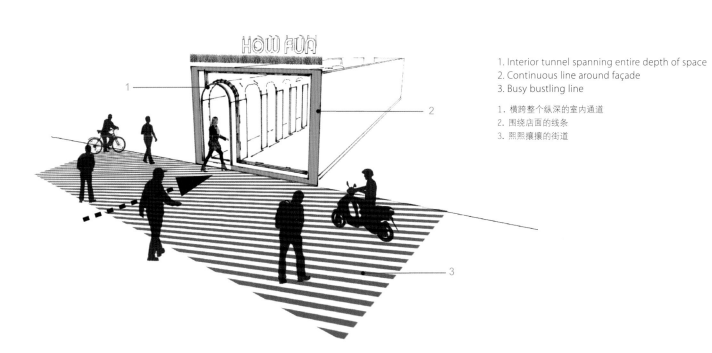

1. Interior tunnel spanning entire depth of space
2. Continuous line around façade
3. Busy bustling line

1. 横跨整个纵深的室内通道
2. 围绕店面的线条
3. 熙熙攘攘的街道

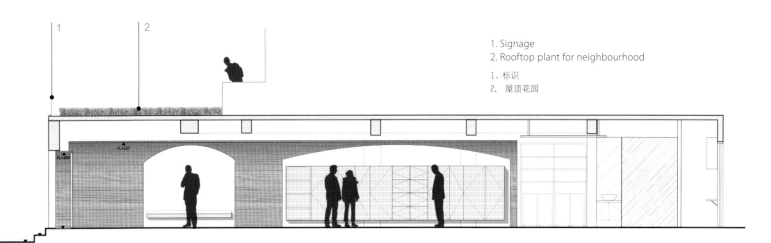

1. Signage
2. Rooftop plant for neighbourhood

1. 标识
2. 屋顶花园

FRESH Beauty Chain Shore

芙雪美容连锁店

Design: Wesley Liu
Location: Chengdu, China
Completion: 2013
Photography: Wesley Liu

店面设计：廖奕权
店铺地点：中国，成都
竣工时间：2013 年
图片摄影：廖奕权

The shop is focused on losing feathers and whitening. So the designers choose white coating as the façade colour. White means pure and clean. This is consistent with the purpose of the beauty shop.
The designers chose the landing door and window to construct the permeability. The entrance lobby is not very light in the night compared with the light 'FRESH' logo and the interior. It is because it is not allowed to install too many lights at the entance, and the designers wanted to have some contrast in daytime and nighttime. So the light of 'FRESH' logo stands out in the dark.

设计师选择了白色外墙，代表纯净和清洁，呼应店铺脱毛与美白的经营内容。
平台式门窗增加了店面的通透性。由于相关规定不允许在入口处安装过多照明装置，夜里的入口大厅不如"Fresh"店面标识明亮。此外，设计师也考虑到营造日夜不同的店面形象，因此选择了在夜间突出以"Fresh"为主的标识。

Deborah Milan Flagship Store

黛博拉米兰旗舰店

Design: Hangar Design Group
Location: Milan, Italy
Completion: 2013
Photography: Hangar Design Group

店面设计：Hangar 设计集团
店铺地点：意大利，米兰
竣工时间：2013 年
图片摄影：Hangar 设计集团

Arranged on one level on forty square meters, the flagship well illustrates the concept created by Hangar Design Group, focused on maximum emphasis for product, which becomes the only protagonist of the entire space. The store outside is candid, amplified by the reflective surfaces of the furniture and the iridescent panels on the walls, which echoes the vibrancy of the colour fading in the neutrality of the variations of white. Standing on the walls, in neutral shades on a bright texture, the sign of the brand, also appears in the outside mirror glass. Reflective material, brushed steel or polished, is also the façade, an explicit call to take care of ourselves and to indulge in a full immersion in the Deborah world of beauty.

这家在同一楼层上占地 40 平方米的旗舰店完美地诠释了 Hangar 设计集团的设计理念，即将设计重点最大程度地集中于产品。这也是贯穿这个店面工程的主题。店面空间十分通透，家具的反光表面和墙上的反光板映衬得室内格外宽敞，与渐变的店面配色遥相呼应。店铺的品牌标识选用中性色调和明亮材质，也出现在店面外的镜面玻璃中。外墙装饰中还使用到了拉丝不锈钢和抛光不锈钢等反光材料，提醒人们关爱自己，充分享受黛博拉品牌打造的美丽世界。

Kyoto Silk

京都丝绸美妆店

Design: Keiichi Hayashi
Location: Kyoto, Japan
Completion: 2009
Photography: Yoshiyuki Hirai

店面设计：早矢仕继一
店铺地点：日本，京都
竣工时间：2009 年
图片摄影：平井良之

Kyoto Silk, a beauty cosmetic shop advocating natural silk ingredient, is located in the center of Kyoto, which is a famous cultural city in Japan. A traditional wooden Machiya of 80 years' history is transformed into a cosmetic shop at downtown Kyoto, with pure white structure hidden in the existing building, symbolizing the rebirth of perfect skin. Designers chose white paint for the wooden frame surrounding the shopfront to create a clean modern ambiance.

京都丝绸是一家坐落于京都市中心的美妆店。京都中心区 80 年历史的传统木造町屋改造而成，强调天然丝绢保养成分的美妆店，洁白净透的新屋造型潜藏在旧屋躯壳里，象征产品让肌肤宛若破茧而出般的新生。设计师将店面四周的木框架漆上一层白色的油漆，营造出一个干净现代化的区域。

TOOL TATTOO, Tattoo & Body Piercing Studio

TOOL TATTOO 文身和身体穿孔工作室

Design: Estudio Vitale
Location: Valencia, Spain
Completion: 2012
Photography: Estudio Vitale

店面设计：比塔莱设计工作室
店铺地点：西班牙，巴伦西亚
竣工时间：2012 年
图片摄影：比塔莱设计工作室

The project connects to the creative and artistic spirit of the company and seeks to normalize the image of this kind of business and guide to everybody.

The business is aimed at all audiences, breaking the negative connotations associated with the sector to create a well-lit and friendly appearance that contrasts with the notion that a tattoo studio should be a dark and saturated aesthetic.

The external wall decorated with motif Old School tattoo from the 40's and hanging tattoo sketches reinforces the creative workshop spirit.

该项目突出了 TOOL TATTOO 品牌具有创造力和艺术精神的核心理念，设计师尝试将文身和身体穿孔的经营内容日常化，并向大众推广。

店面设计方案打破传统观念中与文身文化相关的深沉、神秘的消极形象，打造明亮、友好的店铺面貌，有助于提升大众接受度。

店面外墙使用 40 年代流行的复古文身图案和悬吊的文身草图进行装饰，突出创意的品牌特色。

Sala de Despiece

Sala de Despiece 餐厅

Design: OHLAB / Paloma Hernaiz and Jaime Oliver
Location: Madrid, Spain
Completion: 2013
Photography: Miguel de Guzmán

店面设计：OHLAB 工作室 / 帕洛玛·赫耐兹，杰米·奥利弗
店铺地点：西班牙，马德里
竣工时间：2013 年
图片摄影：米格尔·德·古斯曼

The designers would like to design a very unique and alternative project. The whole façade was covered by a huge white wavy ceramic plate. They used the black character to design a simple interesting graphic work on the plate and the shop window. It involves all the basic information of the shop and makes it easy for people to contact them. The entrance have a part of blue ground which means the ocean. The white sign light box hanging on the shop attracts customers.

本项目中，整个店面外墙被大块的白色波纹瓷板覆盖，与黑色搭配组成有趣的平面图案。其中包含了店铺的基本联系信息，方便消费者进行联络沟通。入口的一块地面是蓝色的，代表海洋。白色的灯箱起到吸引消费者的作用。

goodcals

goodcals 食品店

Design: Hassan Hamdy Architects
Location: Cairo, Egypt
Completion: 2012
Photography: Tarek Mahmoud

店面设计：哈桑·哈姆迪建筑师事务所
店铺地点：埃及，开罗
竣工时间：2012 年
图片摄影：塔里克·穆罕默德

While walking at night on one of the streets of Zamalek district in Cairo goodcals façade catches your eye with its simple minimal lines and catchy fresh colours.

The main aim was to make contemporary façade that reflects the identity of the shop which is fresh healthy and trendy. The shop serves healthy food, and there was a challenge of how to hide the Air Conditioning Units. The idea of putting it in an internal rafter and covering it with a stainless grill enabled the designer to preserve the minimal contemporary style for the façade.

The elements for the façade are simply, the transparent glass framed with an L shape reinforced Iron mesh that is plastered and white painted above it is a stainless grill that covers the air conditioning units at the same time act as a shiny background for the brand signage (goodcals).

夜幕降临后，走在开罗的扎马雷克区街头，设计简约、配色鲜艳的 goodcals 食品店一定会吸引你的注意。

本项目的主要目的是为业主打造一个现代化的店面设计，反映出该品牌新鲜、健康、时尚的形象。设计师面临的一个难题是如何将空调设备隐藏起来。最终选择的方案是将空调纳入外墙结构，覆以不锈钢格栅，在不影响功能的前提下保存了外墙的现代感。

外墙包含的元素十分简单，透明玻璃与刷成白色的 L 形钢筋网架外框，配合覆盖空调设备的不锈钢格子，为品牌标识充当闪亮的背景。

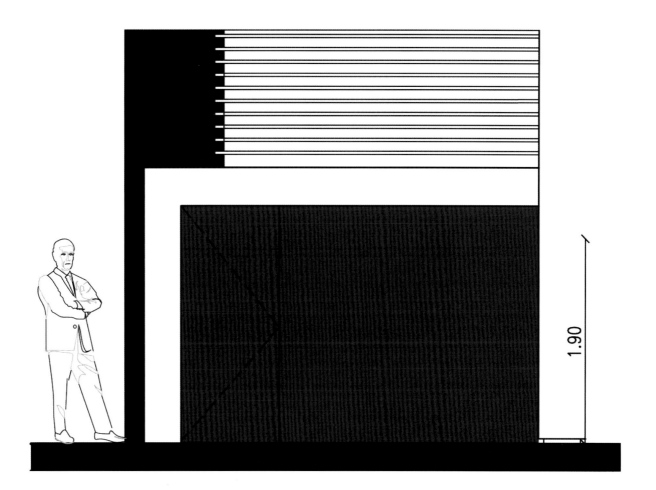

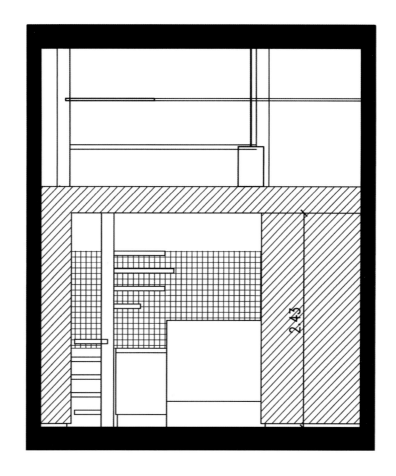

Manish - Arab Cuisine

MANISH 阿拉伯餐厅

Design: ODVO Arquitetura e Urbanismo
Location: Sao paulo, Brazil
Completion: 2011
Photography: Pregnolato e Kusuki estúdio Fotográfico

店面设计：ODVO 建筑设计公司
店铺地点：巴西，圣保罗
竣工时间：2011 年
图片摄影：Pregnolato & Kusuki 摄影工作室

The shopfront is covered with muraxabi concrete shelf in Arabesque. Through a giant screen of hollow patterns, the costumers can enjoy the street scene at a shelter-like position. The space integrates with the surrounding via elements of Arab features and communicates in a subtle yet efficient way. This façade concept of the Manish restaurant pulls all the design elements together to present an impressive shopfront.

带有阿拉伯式花纹的混凝土 muraxabi 框架覆盖了整个立面，保护着内部空间，立面通过无数小框架隔开了室内与室外，通过立面能够看到繁忙街道的景象。庭院、阳台、简洁的楼面，这些元素常常出现在地中海建筑中，在本案中，它们沐浴在透过树木洒下来的阳光下，享受着怡人的清风，一切都显得如此美妙。空间通过能体现阿拉伯文化的元素融入到了周围的环境中。它通过比较含蓄的方式与周围环境形成对话关系并与其共存，这种方法虽不常用，但却是建造这座建筑的基础。

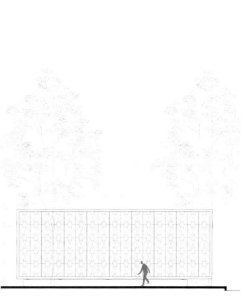

Padarie

Padarie 咖啡店

Design: CRIO Arquiteturas
Location: Porto Alegre, Brazil
Completion: 2013
Photography: Marcelo Donadussi

店面设计：CRIO 建筑师事务所
店铺地点：巴西，阿雷格里港
竣工时间：2013 年
图片摄影：马塞洛·唐纳杜斯

A special mask was applied to the existing façade to provide privacy to the storage room and the office without blocking the sunlight. The idea is very simple, a wood portrait frame that contains a set of coloured metal boards shaped like the wheat spike, the main ingredient of the bread. A wooden pergola completes the façade, limiting the outdoor sitting area. The elements of this project were designed to grant personality and a strong identity to the first Padarie shop in town.

设计师在建筑原有外墙上增加了特殊的遮挡结构，提高储藏室和办公室的隐私性，同时保证足够的光照。设计思想十分简单，木质框架内是一组彩色的麦穗形状的金属板。木质藤架对室外休闲区域起到限制作用，使店面设计更加完整。各个元素的设计充分体现餐厅的风格和个性。

Trattoria Kodama

小玉意式餐厅

Design: SWeeT CO.,ltd, Takeshi Sano
Location: Tokyo, Japan
Completion: 2012
Photography: Nacasa&Partners Inc.

店面设计：SWeeT 有限公司，佐野岳士
店铺地点：日本，东京
竣工时间：2012 年
图片摄影：Nacasa 股份合作有限公司

The designer designed the façade to be covered with wood. It differentiates this building from other urbanized concrete environment. But as the same time, the wood façade also introduces a natural warmth to the environment.

For signage, the designer decided to use the original logo, which was inspired by the shape of Jamon Serrano, and Italian tricolore as a background. It demonstrates the characteristic of this Italian restaurant where serves the aging beef as their special feature.

设计师将店面打造成木质外墙，与其他混凝土建筑区别开来。同时，木质外墙还为店铺增添了自然温馨的氛围。

标识方面，设计师选择使用店铺原有的标识设计。这个设计采用了著名的塞拉诺火腿的轮廓，以意大利三色旗为背景，展示了这家以排酸牛肉为招牌的意大利餐厅的独特魅力。

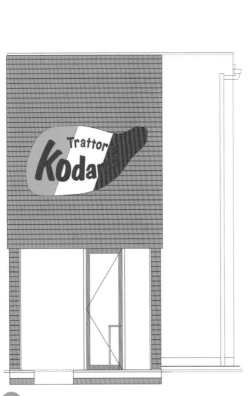

Pizzazaza

Pizzazaza 餐厅

Design: Clifton Leung
Location: Hong Kong, China
Completion: 2012
Photography: Shia Sai Pui

店面设计：梁显智
店铺地点：中国，香港
竣工时间：2011 年
图片摄影：佘世培

Pizzazaza is located in Hong Kong's Tai Hang area, an up-and-coming neighbourhood that offers relaxing dining away from the fast-paced city. The façade plays a vital role in creating a memorable impression on the customers before entering the restaurant. Stylish European folding doors in white frames with glass in yellow shades are used at the entrance to impart a sense of refreshing coziness. Shutters in black are incorporated to further enhance the chic European look and give a dynamic contrast. The folding doors add flexibility to create a more open space. The design creates a translucent effect, resulting in a seamless blend between the interior and the exterior.

The eye-catching spotlights on the top of the entrance create an easy-to-spot and inviting entrance. The overall colour scheme in white and translucent yellow creates a minimalistic backdrop for the rich adornments within.

Pizzazaza 餐厅位于香港大坑，一个发展前景尚好的区域。店内就餐环境轻松，使人远离城市匆忙的快节奏。外墙的独特设计使客人在进入店面之前就感受到餐厅的风格与氛围，留下深刻印象。时尚的欧式折叠门采用白框和黄色玻璃，为入口处增添了清爽的舒适之感。黑色的百叶窗进一步加强了别致的欧洲风情，形成一种动态反差。折叠门增加了空间的灵活感和开阔性。这种半透明的整体效果使店面内外和谐过渡，天衣无缝。

入口上方引人注目的聚光灯打造出一个易于辨识，温馨友好的入口氛围。白色与半透明黄色搭配的整体配色方案营造出简洁的背景，与丰富的装饰物巧妙搭配。

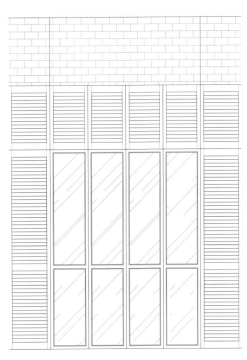

ASK Italian Hertford

德特福德 ASK 意大利餐厅

Design: Gundry & Ducker Architecture
Location: Hertford, UK
Completion: 2011
Photography: Hufton + Crow

店面设计：甘德利＆达克建筑公司
店铺地点：英国，德特福德郡
竣工时间：2011 年
图片摄影：霍夫顿与克劳摄影公司

In this project for ASK Italian, glass is used as the decorative material in the shopfront.
Green interior contrasts with the white façade, contributing to a clean and refreshing dining environment. The brand logo is delivered with a brief bright pallet,

ASK 意大利餐厅为营造一个暖和轻松的就餐环境，外观以玻璃为主要装饰材料。
室内的绿色与店面的白色相呼应，如同青草与白云般清新唯美，打造出回归自然的舒适就餐环境。店铺 logo 采用简洁、明快的颜色，单纯清晰的字体和工艺，易于和过往行人进行信息交流，而且能够给人留下深刻的印象。

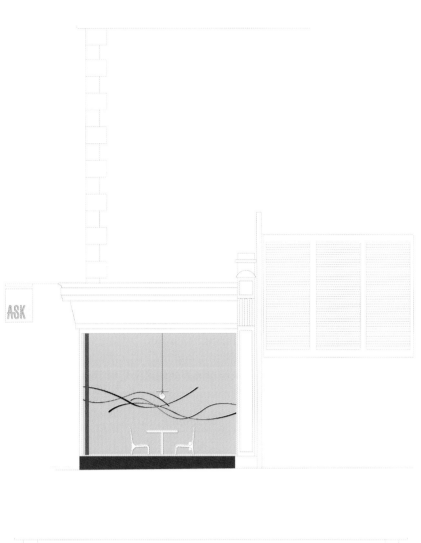
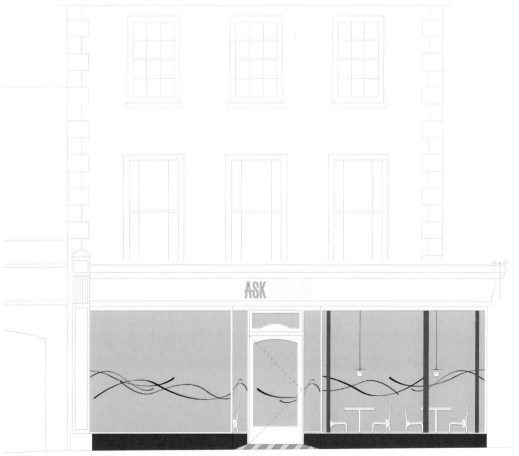

P.S. Restaurant

又及餐厅

Design: Golucci International Design
Location: Beijing, China
Completion: 2012
Photography: Sun Xiangyu

店面设计：Golucci 国际设计公司
店铺地点：中国，北京
竣工时间：2012 年
图片摄影：孙翔宇

Located in Zhongguancun in Beijing, P.S. Restaurant features a naturally relaxing green colour scheme. The refreshing pallet together with natural marble gives those who work around the clock a break they deserve. The temptation to a tasteful meal and rest of mind one get by simply looking at the shopfront is hard to resist.

位于北京中关村的 P.S. Restaurant 又及餐厅，很特别，都是淡淡的绿色，给人清新放松的感觉。唤起人们校园食堂的回忆，柔和的绿色系色彩和天然的大理石如同一个有机的调色盘，提供刚刚踏出校园的年轻学子们心灵加油站，设计师利旭恒希望提供一个闹中取静的幸福空间。光从外面走过就觉得很有食欲，想错过都难。

设计师除厨房、吧台等基本后场之外，所有的外场用餐区域以环境心理学的模式呈现，同时透过窗口静观这纷扰的城市，为不同的人们创造一个属于他们自己的心灵加油站。

PART 8 Works-Catering

Master Food Restaurant

大师美食餐厅

Design: Plotcreative Interior Design Limited
Location: Hong Kong, China
Completion: 2013
Photography: Plotcreative Interior Design Limited

店面设计：柏誉创作策划有限公司
店铺地点：中国，香港
竣工时间：2013 年
图片摄影：柏誉创作策划有限公司

Master Food Restaurant is a local restaurant of minimal decorative style, one of the most popular art trends in recent years. Designers used common materials such as wood panels and metal trims to create a simple relaxing environment to dine in. The sign of the restaurant is shining on the black sign board in night. The shop window and the menu also are lighted by many spot lights.

大师美食餐厅是一家极简主义的临街餐馆。沿用这种当今主流的艺术风格，设计师用木板、地摊、金属包边等简单的材料打造出清心静雅的就餐环境。极简主义外观，尽显静雅气质，用最基本的表现手法来追求最精华的部分，能引发更为深刻的视觉震撼。

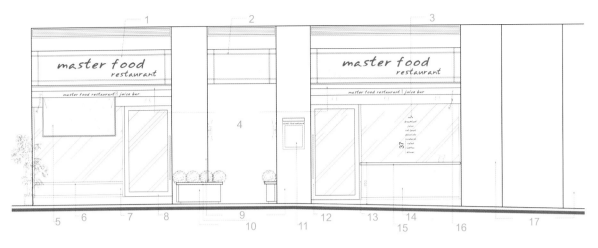

1. 黑白配色标牌
2. 20毫米x20毫米黑色框架，木质表面处理
3. 磨砂表面窗口
4. 黑色店面标识
5. 黑色板材、白色边框的菜单
6. 木质框架
7. 暗灰色金属板
8. 深灰色门框
9. 原有灰色瓷砖
10. 黑色金属表面
11. 黑色板材和木质标识
12. 暗灰色金属门框
13. 木质框架.
14. 暗灰色金属板
15. 木质表面搁架
16. 黑色标识
17. 原有灰色瓷砖

1. Signage board w. black matel & white colour led logo w. S/s frame fin.
2. 20x20mm black frame w. wooden fin.
3. Window sticker w. sand fin.
4. Tent w. black colour and white colour signage
5. Juice bar menu w. black broad & wooden frame fin.
6. Wooden frame.
7. Matel panel w. charcoal grey
8. Matel door frame w. charcoal grey
9. Existing grey tiles
10. Matel fin. with black colour
11. Black board and wooden frame logo w. black colour
12. Matel door frame w. charcoal grey
13. wooden frame.
14. Matel panel w. charcoal grey
15. shelf w. wooden fin.
16. White colour frag and black logo
17. Existing grey tiles

ODESSA Restaurant

奥德萨餐厅

Design: YOD Design Lab
Location: Kiev, Ukraine
Completion: 2013
Photography: Andrey Avdeenko

店面设计：YOD 设计工作室
店铺地点：乌克兰，基辅
竣工时间：2013 年
图片摄影：安德烈·阿夫迪恩科

The atmosphere was formed largely by perception of materials and light. The designers wish to make it eye-catching. The space of the main hall is formed around the island lounge zone decorated with ropes stretching for about 30 km in total. In the evenings, floor lamps directed at the ropes create a cosy lounge atmosphere changing the emotional perception of the place drastically. The play of light can be well seen from the façade attracting a visitor to look inside. The wood sign board with the yellow light seen through the vertical plywood water-resistant. And the swimming and colourful fishes show on the terrace attract the people to come here.

设计师利用材料和灯光的组合营造理想的餐厅氛围。大厅以休闲岛为中心，这里以总长 30 千米的绳索装饰。夜幕降临时，落地灯照射在绳索上，光影交错间构成闲适的气氛，显著改变店面的空间感官效果。这种效果透过店面窗口对行人构成极大的吸引力。店面露台处绘出的五颜六色的鱼充满趣味。

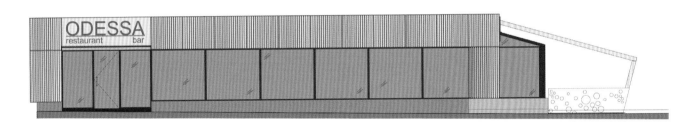
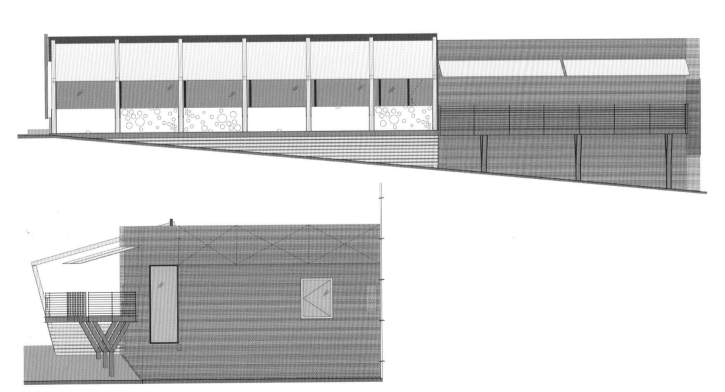

Olivocarne Restaurant, London

伦敦橄榄树餐厅

Design: Pierluigi Piu
Location: London, UK
Completion: 2010
Photography: Pierluigi Piu

店面设计：皮耶路易吉·皮乌
店铺地点：英国，伦敦
竣工时间：2010年
图片摄影：皮耶路易吉·皮乌

Olivo restaurant opened in November 1990 specialising in Sardinian cuisine serving the local Belgravia area.
The access to the restaurant and its appurtenances opens on the aubergine coloured shopfront. The solid wall which previously divided the entrance from the dining room facing the street has been knocked down and replaced by a full height fire resistant glazed divider, so to give a visual opening and more breath to otherwise too compressed spaces.

橄榄树餐厅于1990年11月起营业至今，面向贝尔格莱维亚区，主要特色是撒丁岛风味菜肴。
餐厅店面为深紫色。原本将入口和餐厅分隔开来的墙壁被拆除，取而代之的是一扇落地防火玻璃门。这一改动使原本略显狭小压抑的空间开阔通透起来。

Sawasdelight

Sawasdelight 泰式餐厅

Design: Clifton Leung
Location: Hong Kong, China
Completion: 2012
Photography: Shia Sai Pui

店面设计：梁显智
店铺地点：中国，香港
竣工时间：2012 年
图片摄影：佘世培

The project calls for a challenging design transformation from a fusion Japanese restaurant into a modern Thai spa cuisine. Boasting an iconic lotus-inspired design, the pure white façade was fitted with large floor-to-ceiling frameless windows to bring the outside in, creating a visually open and welcoming entrance passage.

Creative lotus motifs are cleverly incorporated into the overall design. The pure white façade features beautiful blooming lotuses rendered in a simple form. The sense of enlightenment continues into the interior with graphic of floating lotuses subtly infused into the entire space.

Echoing the inviting façade design, a giant glass-fronted kitchen is introduced to enhance the sense of permeability, where diners can witness the culinary performance of the chef. Moreover, semi-translucent lotus graphic against a mirrored backdrop creates an interesting and visually soothing effect, inviting guests to indulge in a rejuvenated gastronomic experience.

本项目中设计师需要将一家日本餐厅改造成现代风格的泰式餐厅，具有一定的难度。纯白色外墙配合落地式无框橱窗构成室内外融合之感，辅以标志性的莲花标识，打造出开阔、温馨的入口通道。

莲花标识被巧妙地融入了店面的整体设计。纯白色的外墙上展示盛开的莲花图案，简洁而富有表现力。这种开悟的氛围延伸至室内，表现为整个空间都使用了的飘浮的莲花图案。设计师还选择了巨大的玻璃厨房加强空间的通透感，并与外墙设计相呼应。就餐者可以在这里观赏着菜肴烹饪的整个过程。此外，镜面上出现的半透明莲花图案妙趣横生，帮助顾客深刻地感受餐厅的精致环境，全身心享受美食体验。

Secession

Secession 餐厅

Design: PROCESS5 DESIGN Noriaki Takeda, Ikuma Yoshizawa, Tatsuya Horii
Location: Wakayama Prefecture, Japan
Completion: 2013
Photography: Stirling Elmendorf Photography Stirling Elmendorf

店面设计：PROCESS5 DESIGN 设计公司，武田纪昭，义泽玖磨，堀井达也
店铺地点：日本，田边市
竣工时间：2013年
图片摄影：斯特林·埃尔门多夫摄影

This is a new construction plan of a restaurant in Tanabe City located in Wakayama Prefecture. The concept was 'Dinner party that turns into art'.

The designers proposed a façade design that two pieces of walls are piled up like a layer and they are related to art in the restaurant that will be filled with pictures of the Viennese Secessionist.

The first piece layer is the wall of the art frame. The second piece of layer is the wall of laminated molding that is used for an art frame. They will appeal to neighbouring people at the clear place. It is intended for people to feel elation eating at the restaurant.

本项目是位于田边市和歌山县的一处新餐厅建设工程，其店面设计构想为"艺术的晚宴"。

设计师提出的外墙设计方案由两部分外墙组成，这两层结构堆叠，与店内装饰一样紧扣维也纳分离派艺术的核心。

第一层外墙结构是外墙的主体。第二层外墙层压成型，形成画框一般的结构。这样的店面设计无疑具有十足的视觉冲击力，也吊足了人们的胃口。

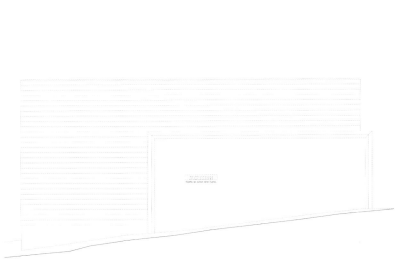

toko-toko

toko-toko 餐厅

Design: KenjiHamagami/switch-lab inc.
Location: Nara, Japan
Completion: 2013
Photography: Kenji Hamagami

店面设计：濱上健兒，switch-lab 股份有限公司
店铺地点：日本，奈良
竣工时间：2013 年
图片摄影：濱上健兒

It is a discerning restaurant finishing barbecued chicken and the ground vegetables using the Japanese native chicken by a charcoal fire. It is the mission of this project to take in the casual nature in the psychogenic dignified expression to control of the sum.

The mix of two colours like sand wall coating agent and binder acrylic resin aqueous type is applied, as the designer painted in Japanese style in the waveform. Sign, such as stone, has given a profound impression as it was aging paint the lamination of acrylic. Purple store curtain can be found in the entrance,

这是一家经营烤鸡和蔬菜的餐厅。本项目的目标是将自然元素引入设计，打造优雅休闲的就餐环境，同时尽可能控制工程开销。

设计师使用了两种颜色，以砂墙涂料和水性丙烯酸树脂的手法呈现出来。还用到了日本风格的波浪图样。石质的店铺标识充满神秘的纹理。入口处是紫色的幕帘。

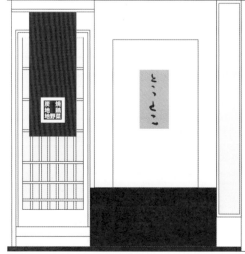

Horiguchi Coffee

堀口咖啡厅

Design: KAMITOPEN Architecture-Design Office co.,ltd.
Location: Tokyo, Japan
Completion: 2013
Photography: Keisuke Miyamoto

店面设计：KAMITOPEN 建筑设计有限公司
店铺地点：日本，东京
竣工时间：2013 年
图片摄影：宫本启介

This project is to redesign Horiguchi Coffee Shops at Setagaya-ku.

This shop design is not complicated. The ground floor is consisted of wood door, window frame and big glass window and the dark sign board. The shop sign 'HORIGUCHI COFFEE' is highlighted by the 5 spot lights at night. The logo on the glass of door echoes the shop sign. On the second floor there have wooden shelves display different cups, plates, bows and containers etc. which identify the shop implicit.

该店面设计项目中设计师需要对世田谷区的堀口咖啡厅进行重新设计。
一楼店面由木门、木质窗框、大玻璃窗和深色标牌组成。店面标识上的五只射灯加强店铺在夜间的受关注度。二楼的木质货架对多种餐具和容器进行展示。

A La Café
爱乐咖啡

Design: Huayang Xiong, Wenwen Xu
Location: Dongguan, China
Completion: 2013

店面设计：熊华阳，徐雯雯
店铺地点：中国，东莞
竣工时间：2013年

Covering an area of 75 square meters, A La Café is a small modern coffee house of simplicity and high efficiency. The original layout was converted to an L-shape interior, to make the most of the limited space. The main purpose was to accommodate the kitchen and the bar indoor, as a relatively generous space is available outside.

Brown paint is used for the wall, as a complementary method of decoration. Lighting focuses on the inner part of the interior to create a bright and welcoming atmosphere. Lighting for the booths, the bar counter, and sofa seating all vary depending on specific needs. The dominant colour of brown, used along with light brown, grey and black, indicates the business category of coffee and visually amplifies the interior.

本案爱乐咖啡厅，面积只有75平方米，加上户外面积简欧风格是由客户要求的，而且小空间比较适合简欧的线条，在造价方面也能控制住施工成本。设计师将原有空间改为L形的平面布局，可以充分使用店铺的利用率，因为爱乐咖啡厅的实际面积比较小。而且户外有很大的区域可以使用，所以室内主要是完成咖啡厅的功能区，如吧台、厨房等。

墙面使用咖啡色的油漆，因为这个项目对品牌的设计没有做包装，所以在墙面使用油漆来装饰。灯光主要集中在咖啡厅的最里端，因为太靠内，灯光亮一些，会显的温馨一些。而且卡座、高吧台、沙发区的灯光都是不同的。基调是咖啡色，符合项目的以经营咖啡为主的市场定位。配色有浅咖啡色、灰色、黑色。因为简单的黑白灰色系，使空间的尺度感放大。

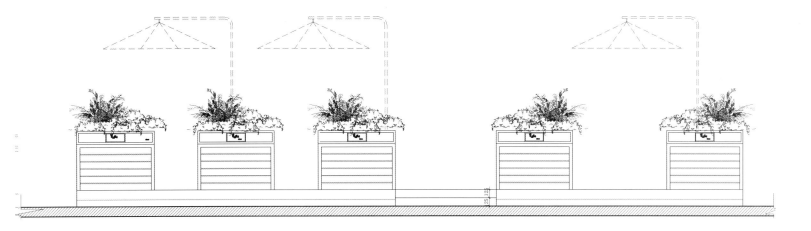

Lindt Cafe Chapel St.

查普尔街林特咖啡馆

Design: Rolf Ockert Design
Location: Melbourne, Australia
Completion: 2012
Photography: Rolf Ockert Design

店面设计：罗尔夫·欧克特设计公司
店铺地点：澳大利亚，墨尔本
竣工时间：2012年
图片摄影：罗尔夫·欧克特设计公司

The design of the Lindt cafes aims to be inviting and classy, somehow the idea of creating a modern world version of a European 'Kaffeehaus' was in the mix.

The materials were selected to evoke this but also of course to represent chocolate. The resulting brown, white and golden colour palette was a far cry from the hitherto trademark Lindt-blue.

Since then the colour scheme developed for the Australian Lindt Cafes has in fact had an impact on the global Lindt ID as more and more elements representing the Lindt brand are now conceived in the brown/gold/white colour range.

林特咖啡馆的设计目标是打造一个温馨而有格调的餐饮环境。与此同时，呈现一个现代版的欧洲"咖啡吧"。

工程选用的材料呼应设计主题，也让人联想到巧克力。最终的棕色、白色和金色的组合搭配虽然明显脱离了林特咖啡厅的蓝色品牌形象，但设计成功而不失个性。

事实上，这家林特咖啡厅的配色方案对其在全球范围内的品牌形象有着极大的影响，越来越多的林特咖啡厅品牌元素呈现出棕、金、白的配色趋势。

1. Façade to remain unchanged
2. New illuminated Lindt logo in lieu of existing 'GLORIA JEANS' signage
3. U/Side shopfront
4. U/Side canopy
5. Tenancy floor level
6. Existing canopy
7. New illuminated blade sign golden Lindt logo on white background

1. 外墙保持不变
2. 全新的林特品牌照明标识代替原来的"格洛里亚牛仔"招牌
3. 店面侧面
4. 遮阳篷侧面
5. 店面地面
6. 原有遮阳篷
7. 白色背景上的金色林特品牌标识

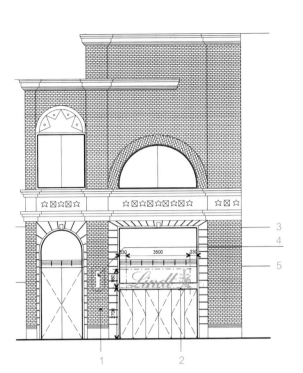
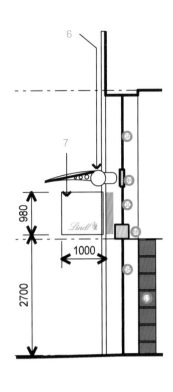

Chocolat Milano

米兰巧克力店

Design: Blast Architects
Location: Parma, Italy
Completion: 2008
Photography: Pietro Savorelli

店面设计：布拉斯特建筑师事务所
项目地点：意大利，帕尔马
竣工时间：2008年
摄影：彼得罗·萨沃雷利

The first Chocolat Milano, in Parma old town, will be guideline for the next openings. Two window modules of this smaller version of Chocolat Milano give onto the square, whilst one looks over the street. The view on the town has been realised through a wise composition of transparent glass and natural iron full surfaces. One of the two windows facing the square is completely covered with iron, exception made for an exhibition niche that runs transversally. The other one, with the entrance, is featured by a one-side only, L-shaped frame, while the rest is transparent and allows a continuous visual interaction between the piazza and the interior. The same solid visual relation is proposed again by the interpretation of the shop window facing onto the Strada Garibaldi. Here the transparency of the all-height glass surface is highlighted by contrast by the presence of two iron bands running parallel to the street, that frame the transversal shop-window destined to products exhibition.

本项目是意大利的帕尔马老城内的第一家米兰巧克力店，将作为范例为品牌发展提供指导作用。店面俯瞰街道。设计师利用透明玻璃和天然铁材料的巧妙搭配呈现帕尔马老城的景色。面向广场的两扇窗户中有一个完全被铁覆盖，只留出一条缝隙。另一扇窗户与入口连在一起，安有L形框架，其余部分为透明玻璃，实现外部广场与内部空间的视觉互动。同样的视觉关系还存在于面向街道的橱窗。落地大窗的通透感与两个与街道平行的铁环形成反差，构成横向的橱窗展示框架。

Les Bébés Café&Bar

贝贝西点西式餐厅 & 杯子蛋糕专卖店

Design: JC Architecture
Location: Taiwan, China
Completion: 2013
Photography: Kyle You

店面设计：JC 建筑设计
店铺地点：中国，台湾
竣工时间：2013 年
图片摄影：游宏祥

The yellow 'folded' box at entrance in pairs of the white showcase, for customers, this sense of delight, surprise and freshness of transformation from the entrance make people delight to come in.

The designers took the box and simplified it down to its structural elements. With 4 different folding steps, 0, 30, 47 and 90 degree angle, the designers folded the box from its original state to a 3D object. The designers created a series of these boxes in these 4 angles so the transition of the exterior of the shop is completed.

入口处黄色"折叠"盒子等新奇、活泼的细节设计对潜在消费者有很强的吸引力。
设计师将包装盒的概念简化为结构元素，应用在店面外墙装饰中。通过不同的步骤，按照 0 度、30 度、47 度和 90 度角进行折叠，可以得到立体的盒子。设计师用这些盒子进行了一系列的外墙改造。

1. Elevation
2. Bar
3. Folded box wall and ceiling
4. Inside
5. Outside

1. 立面
2. 吧台
3. 墙壁和天花板细节
4. 店铺内部
5. 店铺外部

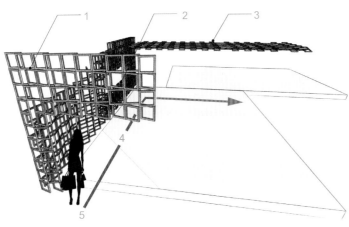

Bolo ao Forno!

Bolo ao Forno！甜品店

Design: Estúdio Jacarandá - arquitetura + design / Luciana Carvalho and Renato Diniz
Location: São Paulo, Brazil
Completion: 2014
Photography: Cezar Kirizawa

店面设计：Estúdio Jacarandá 建筑设计公司 /
卢恰娜·卡瓦略，雷纳托·迪尼兹
店铺地点：巴西，圣保罗
竣工时间：2014 年
图片摄影：切扎尔·克里扎瓦

This is a simple façade to highlight the vertical showcase that enhances the vintage stove and logo 'Bolo ao Forno!' This façade is coated with paint that looks like cake topping and the entrance of the store can be considered an open oven.

The front of the store was designed to be simple, clean and neutral, to contrast with the store and window. The focus is the showcase with the old stove, which works as a support for the products. Materials and elements used on the façade include painting with looks irregular cement; application of unique patterning on the sides of the window; red awning; and illuminated sign with logo 'Bolo ao Forno!'.

在这个简洁的店面设计中，设计师选择突出复古炉子和"Bolo ao Forno 甜品店"的标识。外墙涂料看起来就像蛋糕的装饰配料，店铺入口象征开放的烤箱。

店面是简洁、中性的设计风格，与店铺内部和橱窗的设计形成对比。复古炉子是橱窗的视觉重点，对产品起到补充强化的作用。外墙使用到的建筑材料和建筑元素包括外观像水泥的涂料，橱窗两侧的特殊图案，红色遮阳篷和"Bolo ao Forno 甜品店"的标识。

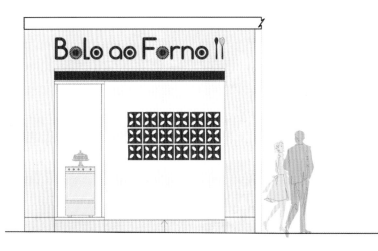
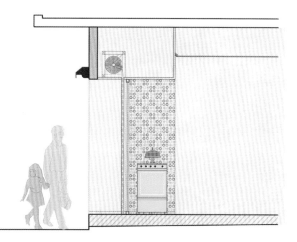

Chocólatras Store

Chocólatras 烘焙店

Design: Studio Cinque, Lisiane Scardoelli, Plexo Design Lab
Location: Porto Alegre, Brazil
Completion: 2013
Photography: Jony Partos

店面设计：桑克工作室，利希安·斯嘉多艾里，Plexo 设计实验室
店铺地点：巴西，阿雷格里港
竣工时间：2013 年
图片摄影：乔尼·巴托斯

Established in the hearth of Moinhos de Vento in a 25m² store, this project started with two main guidelines: zoning and intelligent flows.
The building is preserved by the city, so the existing façade was preserved and complemented with a personalized awning, banners and lights.
The original façade wall of irregular stone ia highlighted with the use of vivid red in. The door and window frames. These elements are designed to grant personality and a strong identity to the Chocólatras store.

Chocólatras 烘焙店位于该城市中心地带，是一个占地 25 平方米的小店。这项店面设计的主要方针是分区和智能分流。由于隶属受保护建筑，原有外墙被保留，在此基础上添加了遮阳篷、横幅和照明装置。
外墙墙面由不规则石材组成。红色的门和窗框与黑色招牌形成反差。这些元素的设置有助于加强空间的风格和个性，突出店铺的品牌形象。

Les Bébés Cupcakery

贝贝西点杯子蛋糕专卖店

Design: J.C. Architecture
Location: Taiwan, China
Completion: 2012
Photography: Kevin Wu

店面设计：柏成设计
店铺地点：中国，台湾
竣工时间：2012 年
图片摄影：吴启民

The designers decided to create an impressive cake package of bakery shop in order to attract customers and promote sale. The 'folded' package concept is largely employed for the interior, the window display as well as takeaway boxes, stirring up the desire for fancy desserts. The designers chose a simple glass window instead of the traditional façade, to provide permeability and a sense of coherence.

There is only one wooden door and a fine window display of exquisite cupcakes on the façade. The classic black and white pallet is highlighted by a touch of creamy yellow, like an irresistable chocolate banana cupcake.

为了营造出一个简洁却不简单的通透空间，好为途人留下深刻印象，设计师决定将蛋糕店塑造成一个蛋糕盒。设计从包装出发，利用折包装盒的概念，由里延伸到建筑物外观，经由空间、橱窗到蛋糕的一系列包装，引导人们对甜点的幻想以及渴望。

门面没有任何花巧和缤纷色彩，选择以一整面落地玻璃来代替传统墙面设计的贝贝杯子蛋糕店，整个外墙立面仅有一扇木门和一个摆放着精致纸杯蛋糕的深框窗口，内部布局一目了然，经典的黑白主调以一抹奶油霜般细腻的鲜黄点亮，就如同一个香甜可口的朱古力香蕉杯子蛋糕，让人垂涎。

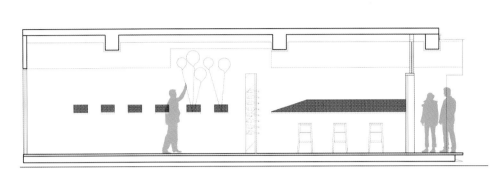
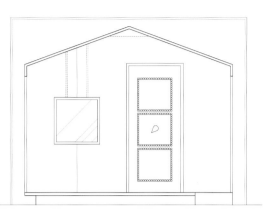

Aoyagi Souhonke Moriyama Shop

青柳森山店

Design: Yukio Hashimoto
Location: Nagoya, Japan
Completion: 2013
Photography: Nacasa & Partners Inc.

店面设计：桥本夕纪夫
店铺地点：日本，名古屋
竣工时间：2013年
图片摄影：Nacasa 摄影合作公司

The exterior wall is finished like rammed earth, as the boxy exterior looks just pulled up from the earth. An organic shape of the entrance and the windows is employed. The interior wall was treated with earthy material, which makes you feel you are in the massive lump of earth.

店面外观采用夯土材质，整个店面看起来像是从土地中抽拉而出的结构。入口和橱窗处采用了有机的设计理念。室内墙壁使用的是外观像土的材料，给人感觉就像置身于一个巨大的土块之中。

1. Sign: Back Channel
2. Outer wall : Stucco
3. Glass
4. Door: Wooden vinyl sheet on steel
5. Outer wall: Stucco

1．标记：反向通道
2．外墙：粉刷
3．玻璃
4．门：钢结构上的木质乙烯基板
5．外墙：粉刷

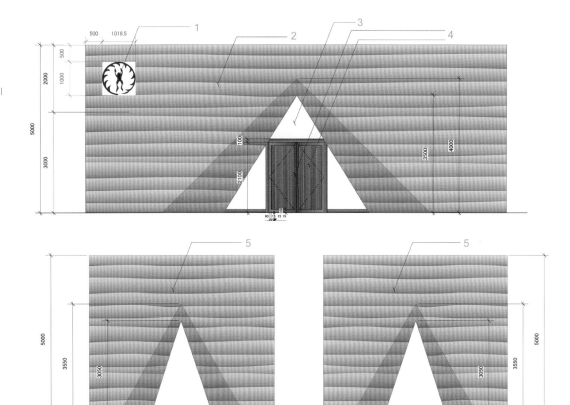

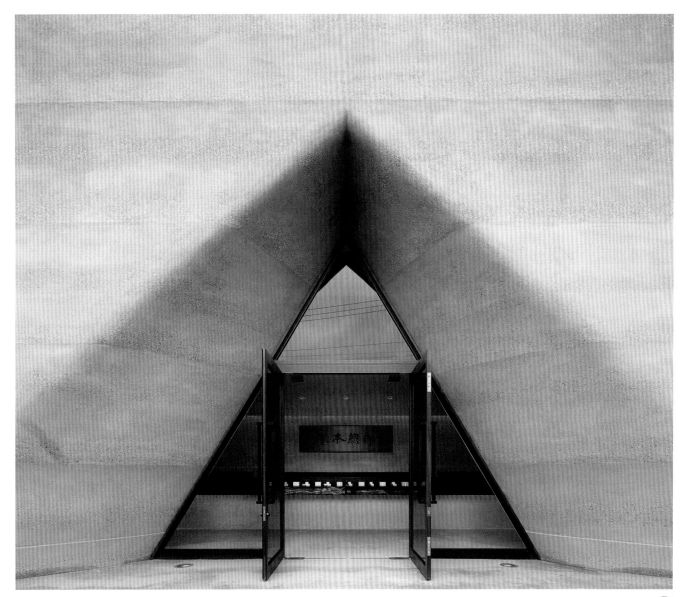

Percimon

Percimon 酸奶店

Design: Sofía Mora – María Velásquez
Location: Barrio Laureles, Medellín, Colombia
Completion: 2012
Photography: Camila Mora

店面设计：索菲亚·莫拉，玛利亚·委拉斯凯兹
店铺地点：哥伦比亚，麦德林
竣工时间：2012 年
图片摄影：卡米拉·莫拉

The spot is located at Jardín Av. at Laureles neighbourhood, in a very green street and with a lot of walking traffic.

The façade was designed to focus the attention on the ice cream shop isolating it from the rest of the original House façade. To give the place a more independent look, the designers used a vertical blind frame. The frame made of metallic white stripes, support the logo and get into de shop guiding the clients to see it and go into the shop.

The design was adjusted to the objective of the brand, a calmed and clean surrounding that leads to rest related to a beach and enjoying environment.

本项目位于查顿街的劳勒莱斯区，这里绿化程度高，步行人流量大。

外墙设计以冰激凌店为中心，将其他的原有外墙阻隔。设计师使用竖直方向的百叶窗框赋予店面更为独立的外观。框架由白色金属条组成，支撑品牌标识，引导顾客进入店内，浏览商品。

整体设计体现品牌目标，呈现清净整洁的环境，提高顾客的舒适度。

Home Cakery

归巢甜品店

Design: SPRS Arquitectura
Location: Oporto, Portugal
Completion: 2013
Photography: Rui Moreira Santos, Espinho

店面设计：SPRS 建筑师事务所
店铺地点：葡萄牙，波尔图
竣工时间：2013 年
图片摄影：鲁伊·莫雷拉

The minimalist blue and white colour scheme creates a natural and comfortable ambiance. This is also employed in the entrance and interior, to achieve a refreshing elegance. Blue decorative pattern against white entrance door adds elegance and personality to the whole design. Mixed with decoration of graphics, the brand logo is pleasant to the eye and also displays an attitude of originality.

蓝白配色的空间，会让人感到自然舒服。在简约风格的门面和空间里大胆运用蓝白配色，让店铺充满清新素雅的美感。尤其在大范围的白色门面上用蓝色作为点缀，让整个设计美观又充满个性。Logo 与几何形状的装饰融为一体，一面让顾客有视觉上美的享受，另一面向人们展示了店铺 logo 别具一格的魅力。

Káxoa

Káxoa 甜品店

Design: SPRS Arquitectura
Location: Oporto, Portugal
Completion: 2012
Photography: Rui Moreira Santos, Espinho

店面设计：SPRS 建筑师事务所
店铺地点：葡萄牙，波尔图
竣工时间：2012 年
图片摄影：鲁伊·莫雷拉

The shop is located in the heart of Maia, a city close to Oporto (Portugal).
The 'cubes' on the show window, an original design, were inspired by the geometry existent on the nature of the elements.
The shop-front is a reflection of the inside shop. The cracked elements are the base to the logo name, representing the same concept of the indoor space. The show window is the main attraction to enter in the fantasy sweet world, so, what better than the proper sugar cubes themselves.

本案位于邻近波尔图的迈亚市中心。
橱窗上的"方糖块"是设计中的一大特色，灵感取自主题元素的几何图形。
店面是店铺内部的缩影。橱窗吸引人们进入一个甜蜜的梦幻世界，"方糖"无疑是最恰当的一种表现元素。

International Giolitti

焦利蒂冰激凌店

Design: NABITO ARCHITECTS (Alessandra Faticanti, Roberto Ferlito)
Location: Istanbul, Turkey
Completion: 2011

店面设计：NABITO 建筑师事务所（亚历山德拉·法提康迪，罗伯托·费里托）
店铺地点：土耳其，伊斯坦布尔
竣工时间：2011 年

Giolitti is the most important hand made ice-cream producer in Rome and in Italy in general. Nabito Arquitectura is proud to design the first Giolitti concept store outside Italy.

The operation is focused on the internationalization of a firm without loosing characteristic of the tradition. The Giolitti front shop expresses individuality but at the same time gives the opportunity of sharing emotions and staying together.

The main idea of the front shop is to give an interpretation of the public space. The architecture shows the will to communicate with the exterior. There are no boundaries between the square and the shop.

A glass mirrored wall on the interior backspace projects the movement of the people inside the shop. And the front table is in contact with public outside. The façade offers a changing system of illumination related with colours of the 'gelato'. The result is a dynamic system of interaction between users and space that modifies the ambience.

焦利蒂冰激凌店是罗马，乃至整个意大利最富盛名的手工冰激凌生产商。NABITO 建筑师事务所光荣受邀为焦利蒂品牌在意大利境外的第一家概念店进行店面设计。

在不损失传统特色的前提下，设计方案着眼企业的国际化层面。焦利蒂冰激凌店的店面设计传递了自身个性，也为消费者提供了分享和交流的空间。

店面的主要设计思想是对公共空间进行阐释，表达与外界沟通的意愿，消除店内店外的界限。

店内的镜面墙反射出人们活动的景象，餐桌也与外界空间相融。店面外墙的照明系统富于变化，与冰激凌的颜色相呼应。这样的动态系统设计在顾客和店面之间构成互动，提升整个店面氛围。

Tempo Patisserie & Café

滕波甜品店

Design: Claudiu Toma, Zeno Ardelean, Norbert Ianko
Location: Timisoara, Romania
Completion: 2014
Photography: Attila Wenczel

店面设计：克劳迪娅·托马，齐诺·阿德琳，诺伯特·伊安科
店铺地点：罗马尼亚，蒂米什瓦拉
竣工时间：2014 年
图片摄影：阿提拉·文策尔

The Tempo shopfront features a red/black colour scheme, emphasizing the sophisticated white logo and echoing with the decorative style of the interior. This establishes a full brand image of the Tempo shop. Bright red panel at the entrance gives direction and enhances the dramatic effect of the façade.

滕波甜品店外部以简约的红、黑为主色，有效地强化了以白色为主的精致 LOGO，同时也巧妙地与室内装饰色彩相呼应，有效地建立起滕波甜品店的整体形象感。入口处竖立的鲜红色面板具有较强的引导作用，更使得店铺的存在感增强。

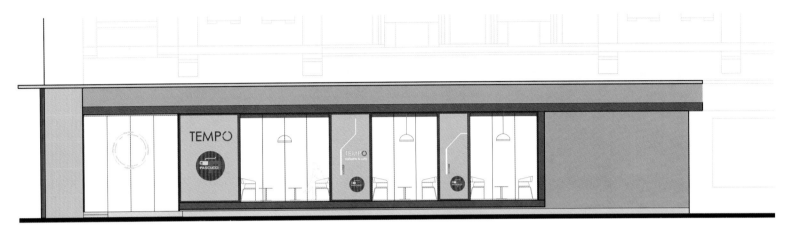

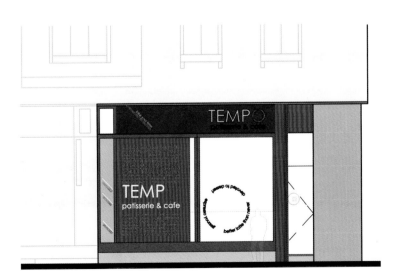

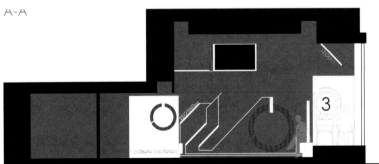

Samba Swirl Frozen Yoghurt

桑巴冰淇淋店

Design: Mizzi Studios
Location: London, UK
Completion: 2013
Photography: Mizzi Studios

店面设计：Mizzi 设计工作室
店铺地点：英国，伦敦
竣工时间：2013 年
图片摄影：Mizzi 设计工作室

This project features lively graphics and LED decoration. The journey starts with pulsating LED light emanating from every plane of the store; walls, floors, ceilings, out and lipping up on to the front fascia. The LED primarily functions as a way-finder system that directs you in and around the store and takes you on a fantastical, colourful, animated, frozen yoghurt voyage. However, the lighting here has more purpose than sole function: the criss-crossed LED blazing routes forms one of the key feature design elements of the store.

店面由充满活力的几何画立面和 LED 照明进行装修装饰。设计师在店铺的墙壁、地板、天花板、店面和横带等各个角度都设置了 LED 照明。它们的主要功能是起引导作用，帮助客人顺利找到出入口，同时营造出梦幻、生动的氛围。交叉的 LED 照明装置还构成店面的主要设计元素。

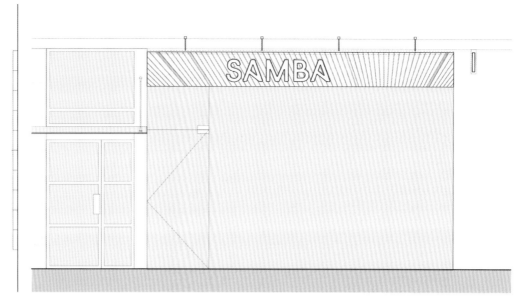
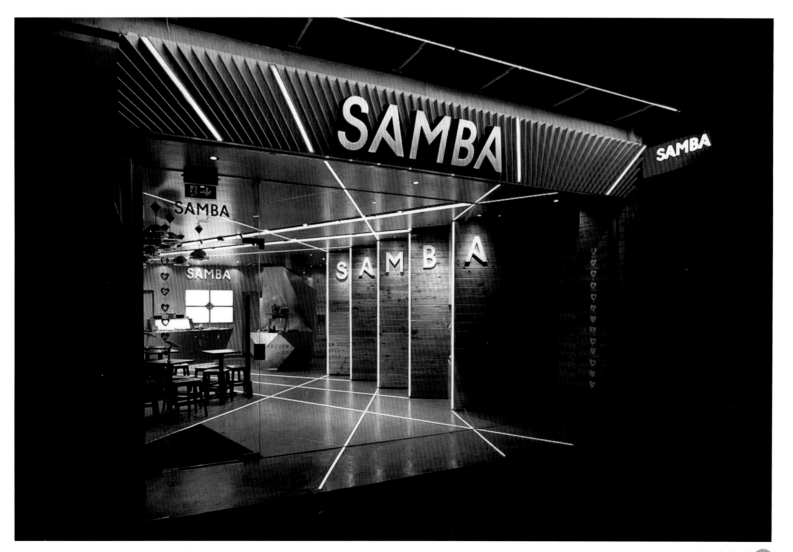

Perles de Chocolat

珍珠巧克力糖果店

Design: SPRS Arquitectura
Location: Oporto, Portugal
Completion: 2011
Photography: Rui Moreira Santos, Espinho

店面设计：SPRS 设计师事务所
店铺地点：葡萄牙，波尔图
竣工时间：2011 年
图片摄影：鲁伊·莫雷拉

This store is located in the heart of the Espinho city, near to Oporto, in the north of Portugal, and it has approximately 70 (seventy) squared meters.

The shop-front of Perles de Chocolat simulates the interior base conceptual design of the interior space, bringing the drop chocolate from the inside. On the shop-front, the dropping chocolate is represented in a white colour and the words are represented in the brown chocolate, like if the words were the chocolate being projected in the white dropping chocolate.

店铺位于葡萄牙北部的埃斯平霍市中心，邻近波尔图，占地约 70 平方米。

这个糖果店的店面设计沿用了室内设计的理念，以流淌的巧克力为表现载体。店面上白色代表浓稠欲滴的巧克力，店名使用巧克力色，好像包裹在流淌的白巧克力中。

Giorno

焦尔诺酒吧

Design: everedge.Inc / Takuma Inoue, Hitoshi Takamura
Location: Osaka, Japan
Completion: 2010
Photography: Seiryo Studio

店面设计：everedge 公司 / 井上诧间，高村蔀
店铺地点：日本，大阪
竣工时间：2010 年
图片摄影：井川工作室

The designers used wood, ceramic tile and clay as the basic material. On the beige clay wall designers created 3 niches which represent Italian colour of green, white and red, with the LED inside, made wine glass frosted to enhance the LED and wine glass represents that this shop sells good wine. The two holes on the top of the beige clay wall exhaust for air ventilation.

设计师选择木材、瓷砖和黏土为基本原料。米色黏土墙上的三个壁龛内置 LED 照明，分别是绿白红的意大利经典配色。红酒杯象征店内出售的保质好酒。考虑到通风的需要，设计师还在米色黏土墙上方留出了两个通风口造型。

El té – Casa de Chás

茶之家

Design: Gustavo Sbardelotto (estudio 30 51), Mariana Bogarin
Location: Porto Alegre, Rio Grande do Sul, Brazil
Completion: 2012
Photography: Marcelo Donadussi

店面设计：古斯塔夫·斯巴德洛托（工作室），
马里亚纳·加林
店铺地点：巴西，南里奥格兰德，阿雷格里港
竣工时间：2012 年
图片摄影：马塞洛·唐纳杜斯

The project concept was born from the immersion in the world of teas. All its colours, textures and aromas were the starting point for creating this environment. Wood was elected as the primary materiality of the project, acting as a neutral base where the colourful herbs are the highlight.

Due to the shop window be visually obstructed by the wall of the shop next door and be quite far from the sidewalk, the store needed a visual attraction that arouse the interest of those who passed through there. For that reason it was sought a synergy between the element of visual communication and architecture.

From the graphical representation of the words 'El TE' chosen as store name, and that literally means 'The Tea', it was developed into a pictogram identification of the tea house that is both visual communication and the main piece of furniture - this goes beyond the scale of a usual sign composing the façade and interior design of the shop.

On the face of 'TÉ', facing the street it was implemented a backlight that functions as an urban lantern, an exciting surprise to those who pass by the store by night. The depth of the letter 'E' on the façade extends beyond the outer limit, penetrating inside the store and acts as the main design element. This element homes the showcase of teas, infusions preparation desk and cashier.

本项目的设计灵感来自于茶的世界。茶的颜色、质感和香味为设计方案提供了最基本的元素。设计师选择木材作为店面的主要材料，在这个中性的基础上突出彩色的茶叶制品的个性与特色。

由于店面橱窗会被相邻店铺的大门遮挡，与人行道也有一定距离，设计师认为应该创造一个吸引力极强的店面以便激起路过行人的好奇心。因此，视觉表达和建筑结构之间的协同配合就显得尤为重要。

店名"El TE"的含义是"茶"，它构成的图形传递了茶馆的形象，很好地起到了视觉吸引的作用。

朝向街道的"TÉ"标识安装了背景灯，为夜色中的店面增添了个性。外墙上字母"E"的高度延伸至店铺内部，结合了茶品、饮品准备区和收银台，成为店内外的主要设计元素。

Green Zebra Grocery

绿色斑马杂货店

Design: LRS Architects, Inc.
Location: Oregon, USA
Completion: 2013
Photography: Sally Painter Photography

店面设计：LRS 建筑设计有限公司
店铺地点：美国，俄勒冈州
竣工时间：2013 年
图片摄影：莎莉·佩因特摄影

Green Zebra is very small for a grocery store, so in order to compete with larger food retailers and other convenience stores, Green Zebra has to be memorable, visually attractive, easy to navigate, have good signage, and offer great products for its customers. Environmentally friendly wood panels are employed, together with the Green Zebra image and logo, to achieve a natural, friendly shopping experience. The different kinds of materials, active font design and generous illustrations attract many people.

绿色斑马是一家很小的杂货店，为了与更大的食品零售商以及其他便利商店竞争，店面设计需要吸引人，令人印象深刻，易于辨识，指示清晰，并且提供性价比高的商品。店面穿插的是绿色环保木质装修材料，与绿颜色的斑马、logo 相呼应，给人亲近自然的舒适感。多种材质的搭配、活泼的字体设计和丰富的插图有助于这一目标的实现。

3M store

3M 概念店

Design: Torafu Architects
Location: Omotesando, Tokyo, Japan
Completion: 2010
Photography: Daici Ano

店面设计：Torafu 建筑设计公司
店铺地点：东京，表参道
竣工时间：2010 年
图片摄影：阿野太一

The maker of the Post-it Notes and DI-NOC Film sheets, 3M, opened its first concept store in Omotesando, Tokyo for a limited time. The interior of the 3M store feels like a walk-in catalogue almost entirely made using 3M materials and goods to further promote awareness of the brand's product line-up known for its variety and technical innovations.

The granite-like tiles at the entrance, inspired by Omotesando's distinctive pavement pattern, gradually turn into a multicoloured surface as visitors move to the back of the store. The checkered pattern then scatters onto the columns and walls, making its way up to the ceiling where it blends with the suspended lights, mitigating the bareness of beams and pillars in the overhead space.

3M 是便利贴和特耐贴膜的设计者。其第一家概念店在东京的表参道限时向大众开放。走进 3M 概念店，就像走进了一本目录，里面收录的几乎都是 3M 生产设计的材料和产品。这一设计使得品牌产品多样、技术创新的特点得到进一步强化，也使整个品牌形象更加深入人心。

入口处仿花岗岩瓦片的设计灵感取自表参道独特的路面纹理，移步到室内，可以发现它渐变成多色。格子图案随之分散到柱子和墙壁上，并延伸到棚顶，与屋顶的吊灯融为一体，显著地减轻顶部空间里横梁和支柱的裸露感。

PART 8 Works-Grocery

La Despensa Bosques

La Despensa Bosques 超市

Design: BXH Arquitectura, Daniel Vargas, Luis Enrique Guillén, José Vargas
Location: Mexico City, Mexico
Completion: 2013
Photography: Jair Navarrete

店面设计：BXH建筑师事务所 / 丹尼尔·瓦尔加斯，路易斯·恩里克·纪伦，何塞·瓦尔加斯
店铺地点：墨西哥，墨西哥城
竣工时间：2013年
图片摄影：雅伊尔·纳瓦雷特

In this project, the sober selection of colours and materials applied to the space, generate an intense contrast with the products, highlighting them and allowing them to act as the main characters of the design. The dark colour scheme gives the impression of mystery, and the idea of finding certain items in a space like this is intriguing itself. The extra-large logo at the entrance and basic information about the store give direction and are visually attractive.

空间配色和建筑材料的选择和店内产品构成了强烈的对比，使后者的特性得到强化和突出，成为店铺中的主体。黑色的店面设计，让人感觉很神秘。你或许会发现穿黑衣服的人比穿其他颜色衣服的人显得更强壮、更具有攻击性呢？对于黑色门头的店面设计上，给顾客的感觉是，希望能在"黑色的店内"找到一件自己喜爱的物品。就像在黑夜中，我们总盼望着冉冉红日一样。Logo超大的字体，给顾客留下深刻的印象。旁边有清晰的店铺网址和电话，更能满足顾客的各方面需求。

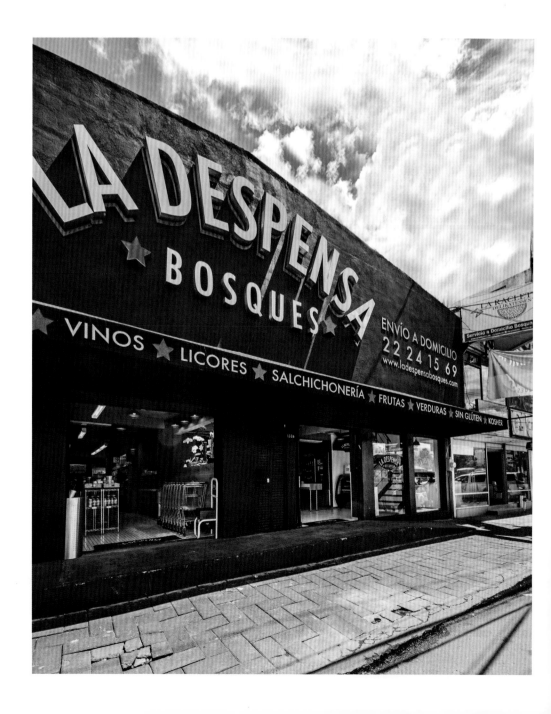

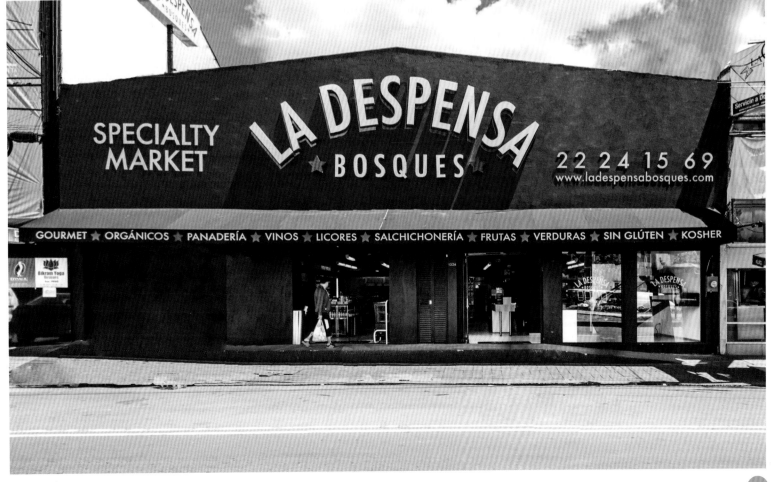

Farm Direct

"农场直达"水培菜零售店

Design: Wesley Liu
Location: Hong Kong, China
Completion: 2014
Photography: Wesley Liu

店面设计：廖奕权
店铺地点：中国，香港
竣工时间：2014 年
图片摄影：廖奕权

This is Farm Direct's first concept store. The store's principal business is in the sales of 100% Hong Kong locally produced hydroponic vegetables. The design concept is about using the cheapest and the most primitive materials to create a European styled store. The red bricks forming the bright exterior are overlaid with a brushed white finish along with a sprayed green vignette, which is a reflection of the fresh produce available in store.
The designers do VI for the vegetable only. The lights are like the roots of the plants. It is like you stand under the ground, and you can see the roots of vegetable growing. The feel is very fantastic.

本项目是"农场直达"水培鲜菜的首家品牌形象店。店内出售百分百香港本土出产的水培蔬菜。店铺的设计是用最原始、最便宜的建筑材料打造一个欧式店面。构成耀眼外墙的红砖结构上涂刷了白色，配合渐变的绿色喷漆效果象征店内新鲜的果蔬产品。
屋顶的照明装置就像植物的根部，置身其中仿佛身处地下，似乎可以观察到植物的生长。这为店铺空间增加了奇妙的色彩。

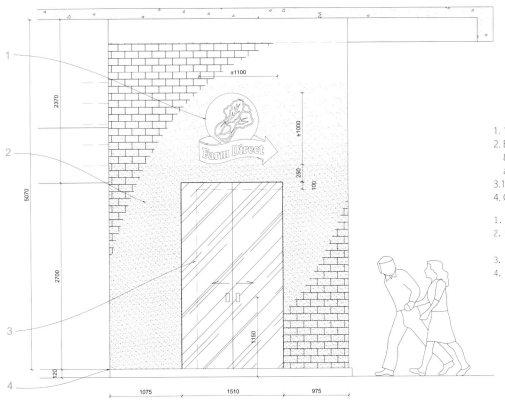

1. 120mm THK. Illuminated acrylic LOGO
2. Brick wall with gradual spray paint (Underlay: dulux 10BB 83/014 drifting white)(Gradual layer: dulux 90GY 216/354 amazon leaf)
3. 12mm THK crystal clear tempered glass
4. Cement floor with volcano ash composite

1. 120毫米厚亚克力标识
2. 渐变涂料墙壁（底层：多乐士10BB 83/014 飘然白）
   （渐变层：多乐士90GY 216/354 亚马逊绿）
3. 12毫米厚全通透钢化玻璃
4. 火山灰复合水泥地面

Läderach Chocolatler Suisse

Läderach 瑞士巧克力店

Design: Studio KMJ (Designer Kurt Merki Jr.)
Location: Baden Baden, Germany
Completion: 2012
Photography: studio KMJ

店面设计：小库尔特·莫奇 / KMJ 设计工作室
店铺地点：德国，巴登巴登
竣工时间：2012 年
图片摄影：KMJ 设计工作室

The uniqueness of Läderach is the fresh chocolate and the design challenge for this project was to create a space that is warm and fresh at the same time. The association of chocolate and brown colour and a couple of classic design elements is used in the storefront, creating a space which has the classic warm feeling, but in addition is presented in a new freshness.

新鲜制作的巧克力是本店的特色产品。本项目中最大的挑战是打造出一个既温馨又清新的商业空间。设计中使用了代表巧克力的棕色和一些经典的设计元素，以令人耳目一新的独特方式创造经典而又温馨的氛围。

Casa Turia

Casa Turia 红酒店

Design: CuldeSac™
Location: Valencia, Spain
Completion: 2013
Photography: CuldeSac™

店面设计：CuldeSac™ 设计公司
店铺地点：西班牙，瓦伦西亚
竣工时间：2013 年
图片摄影：CuldeSac™ 设计公司

Located in the heart of Valencia is the recently opened Casa Turia pop-up store, which celebrates the relaunching of bottled Turia Marzen and generates an enrapturing experience that immerses visitors in the world of the brand. From the outside, Casa Turia is particularly visible and striking, given that its structure extends beyond the entrance windows. The window display is a dynamic device that uses movable modules –made from bottles– in the guise of traditional Valencian latticework.

Casa Turia 红酒店位于瓦伦西亚市中心，是为了纪念瓶装 Turia Marzen 红酒重新投放市场而开设的，这里被打造成一个美妙的红酒世界。
店面外观精致、醒目。传统风格的瓦伦西亚格子窗框架内，橱窗展示的是一个由可移动的瓶子单元构成的动态机制。

Enoteca E

Enoteca E 红酒店

Design: act romegialli, Gianmatteo Romegialli, Angela Maria Romegialli, Erika Gaggia architects
Collaborator: Paolo De Meo
Location: Chiuro, Italy
Completion: 2011
Photography: Marcello Mariana

店面设计：act romegialli 设计工作室，马雷吉尼·罗梅吉亚利，安吉拉·玛利亚·罗梅吉亚利，埃里卡·加吉亚建筑师事务所
合作伙伴：保罗·德·梅奥
店铺地点：意大利，基乌罗
竣工时间：2011 年
图片摄影：马塞洛·马里亚纳

The shop of Enoteca E is part of a bigger main building used as a warehouse and it is situated on a traffic road. In the concrete floor there are wood partitions to delimit different defined areas.
The shopfront with the idea to create a distinctive place with few, simple materials such as larch, concrete, iron. A 'carpet' of wooden planks marks the entrance and extends outside to invite the customer. The street front is characterized by a wide window with overhanging wooden elements to create a materic frame which go with the whole project style. It could show the goods fully.
The shop sign is made of brusche steel. It takes a slender simple font, simple and elegant, without too much complex dressing. A tree and a wooden bench are collocated near the entrance to create a pleasant way to the shop.

店面位于一个主要用作仓库的大楼之内，门口是交通繁忙的公路。混凝土楼板之间使用木质隔板区分不同功能的区域。
设计师意欲用最少、最简单的材料打造出一个风格独特的店面。除了基础的木材、混凝土和钢铁，木板拼成的"地毯"将出现在店铺的入口处，并向外延伸欢迎来客。开阔的橱窗配合天花板悬挂的木质装饰为整个店面工程定下了基调，也便于商品的展示。
店面标识用亚光不锈钢制成。造型简约而不失优雅。入口处树木和长椅的设计更是增加了温馨、惬意的感觉。

Drink Shop Harpf

意大利鲁内克饮料店

Design: Monovolume Architecture + Design / Raumstory
Coworkers: Luca di Censo, Sergio Aguado Hernàndez
Location: Bruneck, Italy
Completion: 2012
Photography: Jürgen Eheim

店面设计：Monovolume 建筑公司，Raumstory 设计公司
设计团队：卢卡·迪琴索，塞尔吉奥·阿瓜多·赫尔南德斯
店铺地点：意大利，鲁内克
竣工时间：2012 年
图片摄影：尤尔根·伊罕

The entrance area was transformed into a traditional corner shop where the passion for detail appears; while in the rear area old architectural treasures have been combined with modern ideas, as defined in 'Rough Luxe'. Already used and worn out furnitre were artfully staged, giving the impressive ruins an extraordinary atmosphere. The Harpf-and-Friends-corner is the connection between food and beverage area where a library is situated, which invites you to take a place. Photos from the past tell the story of the company founded in 1919.

店铺入口处被改造成了传统的街角小店，设计师在这里加入了许多细节：围绕"粗糙的奢华"这一主题，将古旧建筑元素与现代设计理念相结合，经过合理的设计和摆放，破旧的家具绽放出别样的风采。这个转角小店提供食品和饮品，精心布置的老照片讲述过去的故事。

Charles Smith Wines

查尔斯·史密斯酒庄

Design: Olson Kundig Architects
Location: Walla Walla, Washington, USA
Completion: 2011
Photography: Olson Kundig Architects

店面设计：奥尔森·昆汀建筑师事务所
店铺地点：美国，瓦拉瓦拉
竣工时间：2011 年
图片摄影：奥尔森·昆汀建筑师事务所

Charles Smith Wines is located in downtown Walla Walla in the former Johnson Auto Electric building, constructed in 1917. The shell of the building—with original brick walls, wood trusses and a concrete floor—received minor structural updates but was otherwise left raw. The team highlighted the automotive history of the building by replacing garage doors with two custom, hand-cranked pivot doors that completely open the space to the street and form an awning for outdoor seating.

查尔斯·史密斯酒庄位于瓦拉瓦拉市中心的一栋大楼内。大楼建于 1917 年，曾经是约翰逊汽车电器公司所在位置。大楼保留了最初的砖外墙、木桁架和混凝土地面，不需要太多的结构改造，但另一方面略显粗糙。本项目的设计团队选择突出大楼有关汽车行业的历史，将两扇车库大门改造成手动枢轴门，形成适合户外座椅的遮阳篷。

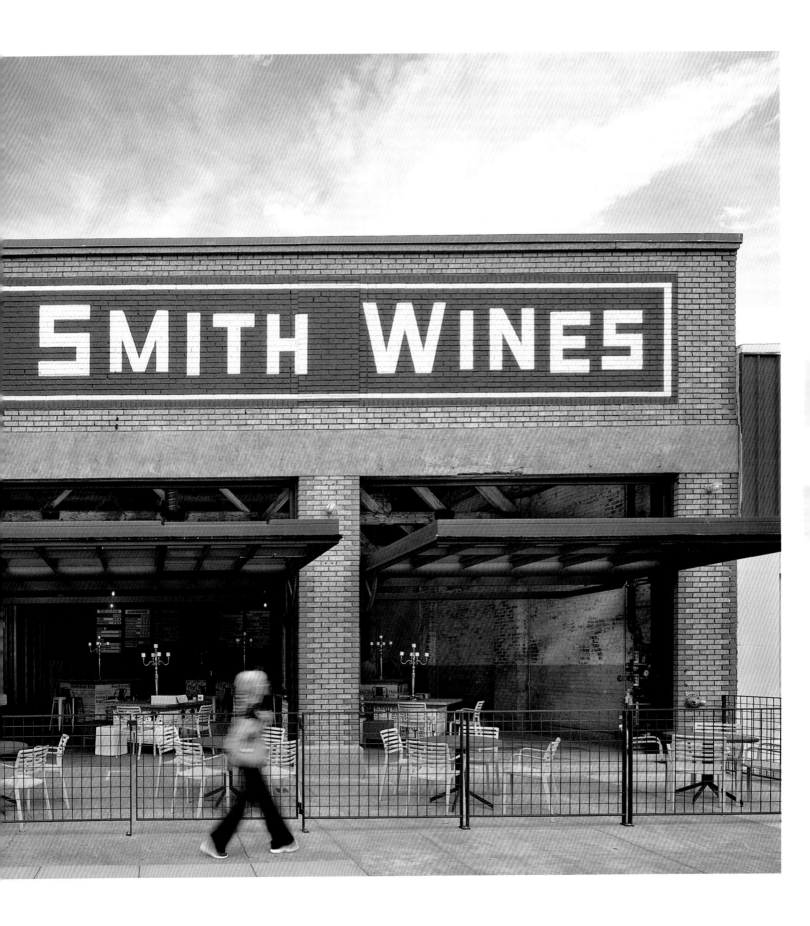

Ligier

利吉尔红酒店

Design: Aarquitectura y Diseño
Location: Buenos Aires, Argentina
Completion: 2013
Photography: Julio Masri

店面设计：构架与设计工作室
店铺地点：阿根廷，布宜诺斯艾利斯
竣工时间：2013年
图片摄影：胡里奥·马斯里

Aarquitectura y Diseño designed the brand image Ligier basing on the wine stores in southern France in the early twentieth century, where attention was personalized and sophisticated. The designers conducted a research work to create a boutique wine and spirits shop that recreates the past, using a classical and ornate aesthetic built mostly in oak, with tile floors and soft lighting.

Large boards on the façade are used to communicate products and advertisements of the moment. The giant glass window enhances the sense of permeability, and it also displays the products well.

The sign is yellow which is a highly visible colour to attract eyes. The yellow illuminated signs to pedestrian and vehicular level help identify the store from anywhere.

利吉尔红酒店的品牌形象设计以20世纪初法国南部风格独特的红酒店为灵感。经过详尽的调查，设计师选择用橡木、瓷砖地面和柔和的灯光构建一个经典而华丽的老式精品红酒店。
外墙上使用大型广告板对产品和店铺起到宣传作用，巨大的玻璃橱窗通透明亮，使商品得到充分的展示。
黄色的店铺标识可见度高，引人注意。

El Mundo del Vino

"世界酒庄"

Design: Droguett A&A Ltda.
Location: Santiago, Chile
Completion: 2010
Photography: Marcos Mendizábal

店面设计：德罗格特有限公司
店铺地点：智利，圣地亚哥
竣工时间：2010 年
图片摄影：马科斯·蒙迪扎巴尔

El Mundo del Vino is the largest wine store and wine distributor in Chile and one of South America. It has several stores in the country and the project of Isidora 3000 building was the relocation of the flagship store.

One of the first drivers of the assignment was El Mundo del Vino's claim: A world of passion for wine, considering that all the work should be done with passion and dedication. Red was the colour chosen for this storefront. Red is the colour for wine and for passion. A red laquered 10x10 cms grid, that remembers the niches in a wine cellar, is the main architectural element of this façade that covers half the lenght of the façade. This grid partially reveals what is happening inside, and gives the visual first impact that the store wanted.

The grid works as a whole with the main entrance doors and with the store enlighted signage. On the other hand, the rest of the storefront is transparent glass, with floor to ceiling wine graphics using the same colours. This is an interesting contrast between two sides with different degrees of transparency.

"世界酒庄"是智利最大的红酒储存商和经销商，也是南美最大的红酒经销商之一。"世界酒庄"在全智利有多家分店，这个名为伊西多拉3000的项目是对旗舰店的搬迁重建。"世界酒庄"的特色是经销世界所有产区的红酒。在这里还可以找到种类丰富的香槟、气泡葡萄酒、烈酒和威士忌等。主要店面中有三个位于圣地亚哥的高档商业区——拉斯孔德斯区。工程的一个主要目标是体现"世界酒庄"的宗旨：打造一个充满激情的红酒世界。因此店面选择了红色为主色调，代表红酒和激情。规格为10厘米×10厘米的红色网格象征酒窖，充当整个店面的主要建筑元素，长度接近店面的一半。这一设计不仅引人联系，还能达到视觉冲击的效果。

红色网格与店面大门、店铺标识协调统一。店面的其他部分使用了透明玻璃，从上到下都是红酒的图样，也使用了相同的颜色。店面两侧建筑材料透明度的不同形成了有趣的反差和对比。

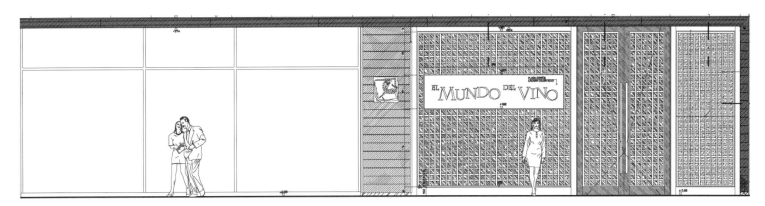

The Whisky Shop

威士忌酒庄

Design: gpstudio
Location: London, UK
Completion: 2013
Photography: John Adrian

店面设计：GP 工作室
店铺地点：英国，伦敦
竣工时间：2013 年
图片摄影：约翰·艾德里安

The Whisky Shop wanted to expand in London and create a flagship store in the heart of London's Piccadilly, opposite The Ritz. The façade of store is designed to educate and inform the consumer as well as convey a sense of the skill and craftsmanship involved in the creation of whisky.

After the designers took the rich heritage and expertise of the product into consideration, they resolved to represent their appreciation of the fine whisky in the façade design. The elegant and classic façade design, deep colour wood texture, just like the colour of whisky, combined with the transparent glass curtain included informs consumers the importance of the ageing process, and how this impacts on the sensory experiences of taste and smell. The 18th-century-style façade with luxury and nobility explains the process of making whisky, such as how the flavours are created by the craftsmen.

威士忌酒庄品牌向伦敦扩张，开设旗舰店，地址选在伦敦的中心——皮卡迪利大街——正对着著名的丽兹酒店。店面设计为消费者提供有关威士忌的知识和常识，鼓励人们欣赏威士忌制作过程中投入的技巧和工艺。

设计师将深厚的威士忌文化遗产和生产工艺纳入设计之中，将对高档威士忌酒的欣赏纳入店面设计方案中。经典的外墙设计和布局、深色的木材、宽大的玻璃橱窗让人联想起威士忌的制作过程，强化感官的冲击力。典雅高贵的 18 世纪风格外墙象征优质威士忌历经多年的制作过程。

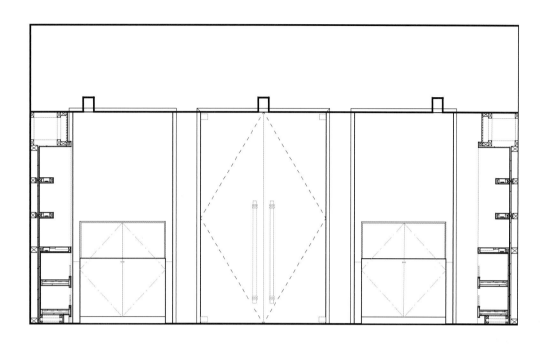
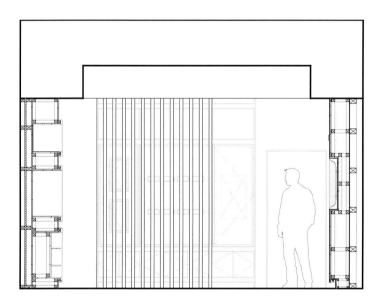
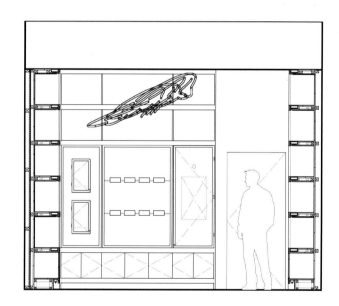

151E

151E 茶铺

Design: PLANNING ES (Shinta Egashira)
Location: Fukuoka, Japan
Completion: 2013
Photography: Hiroshi Mizusaki, Ikunovi Yamamoto

店面设计：ES 设计公司（江头真太）
店铺地点：日本，福冈
竣工时间：2013 年
图片摄影：水崎浩，山本育宪

This project was started with the aim of conveying the charm of Kyushu tea. Unlike other tea shops, this is a place where one can browse freely and casually. The concept behind the interior design was 'a design that gives a feeling of simplicity, warmth and tenderness using raw materials to the maximum.' It was chosen in order to convey the feeling of the raw materials used in making tea.

The entire façade was made of glass, allowing the whole shop to be seen from the entrance. The eaves of the shop, to which the shop sign is attached, are made of basswood. Rolled sheet copper was used on the rafters to heighten the sense of fine quality. The copper will oxide with time, bringing out a unique ambiance. An 'open' signboard was also designed and installed to give a sense of dynamism to the shop.

Traditional Japanese materials and methods were used in the design with modern perspectives in mind. The result is a simple, no-frills shop where one can truly appreciate the raw materials used.

这项工程是为了彰显弘扬九州茶的魅力而开展的。与其他茶铺不同，这里的氛围定位较为轻松随意。室内装潢的设计理念是"利用原材料最大程度地打造简洁、温馨、精妙的环境"。这样也是为了更好地表现茶道中对天然材料的使用。

店面的整个外墙都是玻璃材料，客人在入口处便可以将店内的情况尽收眼底。屋檐上挂着椴木制成的茶铺品牌标识。椽子处使用冷轧铜板以加强做工精细之感。黄铜会逐渐氧化，留下时间的印记，形成独特的质感。"营业中"指示牌的设计和安装方式也透露出动感与活力。

设计过程中不仅沿用了传统日式建材和施工技法，也考虑到了现代生活的习惯与需求。这个简洁、清新的店面设计帮助人们全身心地感受这些原材料的灵与美。

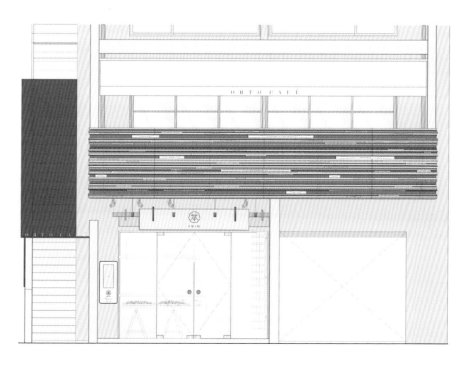

PART 8 Works-Grocery

ARMAZEM

ARMAZEM 食品店

Design: ODVO Arquitetura e Urbanismo, Omar Dalank, Victor Castro, Carol Kaphan Zullo
Location: São Paulo, SP, Brazil
Completion: 2012
Photography: Pregnolato e Kusuki estúdio Fotográfico

店面设计：ODVO 建筑设计公司，奥马尔·达兰克，维克多·卡斯特罗，卡罗尔·卡芬·祖洛
店铺地点：巴西，圣保罗坎波贝洛
竣工时间：2012 年
图片摄影：普雷尼奥拉托与玖月摄影工作室

A place to buy wines and imported food in a comfortable and intimate space to enjoy, learn and enjoy the best that our taste buds can taste. A simple design and sleek material was chosen for the storefront.

In order to maintain detached house built to the neighbourhood, interventions on the façade of its volume were measured, restricting itself to the change of the openings and the installation of a striking metallic blue awning that frames the window, reinforcing the importance of the ground and its mild compared with its neighbours. No gates across the front is proposed, although a small square was built for public use.

这是一家环境舒适，充满私密性的店铺，供应高档葡萄酒、进口食品，是享受美味的好去处。店面设计采用了简约的设计和时尚的选材。

为了保持店铺的独立建筑与周围环境的和谐，设计师预先对外墙改造的施工规模进行了规划和测量，将设计工程简化为门窗的改造和加设醒目的金属蓝色雨搭，既突出了地面的重要性，又加强了店铺与周围环境之间的对比。店面前侧没有设计大门，但设置了一个小型广场，充分满足了公共使用的需求。

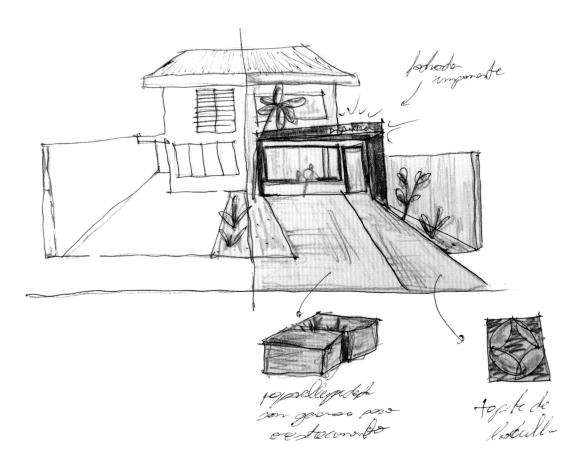

Nu Skin

如新个人保养品店

Design: Green Room Design Team
Location: London, UK
Completion: 2012
Photography: Chris Baynham

店面设计：绿房间设计团队
店铺地点：英国，伦敦
竣工时间：2012 年
图片摄影：克里斯·贝恩汉姆

Taking inspiration from Nu Skin global brand design, the designer used a pure palette of wood, glass and stainless steel, plus various furnishings and integrated AV equipment to provide a welcoming, professional Nu Skin brand experience.
The façade consists of white frame and glass window, presenting a strong sense of depth. The position of entrance and illuminated sign box increase visibility and attract attention.

本项目的设计灵感来自如新集团全球品牌形象，设计师利用木材、玻璃和不锈钢的简单搭配，结合多种样式的家具和视听设备打造温馨、专业的如新品牌体验。
店面由白色框架与玻璃构成，具有较强的立体感，入口处为凹陷的造型，富于个性。带有店铺的名称和 LOGO 的灯箱向外延伸，凸显在整个建筑立面之外，起到良好的引导作用。

De Natural, Natural Health Products Store

De Natural 健康食品店

Design: Estudio Vitale
Location: Valencia, Spain
Completion: 2012
Photography: Estudio Vitale

店面设计：比塔莱设计工作室
店铺地点：西班牙，瓦伦西亚
竣工时间：2012 年
图片摄影：比塔莱设计工作室

'Experiencing nature' is the creative concept that Estudio Vitale has developed for the flagship store of a new natural products franchise. The store goal is to pass health care through natural food brand experience.

The shopfront design follows a minimalist style, keeping harmony with the surrounding. Illuminated logo at the entrance is enhanced by the white façade structure. Canopy over the glass façade provides basic solar protection and serves as a decorative element as well. The overall design effectively emphasizes on the product concept.

"体验自然"是比塔莱设计工作室为旗舰店的全新天然产品而特别设计的概念。目的是让顾客通过品牌消费体验，感知并接受健康养生的理念。

店面整体设计采用了极简的设计风格，保持了与周边环境的和谐性。品牌的标识设置在入口处最显要的位置，被洁白色的墙面衬托得更加醒目。玻璃墙面外部安装有简单的遮阳设施，同时也成为外墙上较为时尚的装饰物。总之，店面的设计有效地强化了产品的品牌理念。

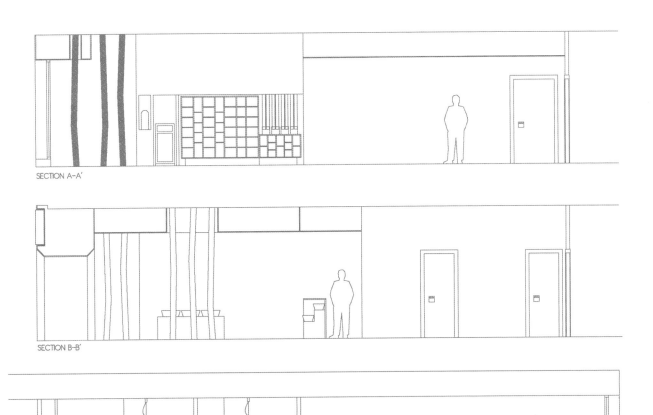

SECTION A-A'

SECTION B-B'

Empório Nanak

信徒大市场

Design: Fernando Maculan and Mariza Machado Coelho (MACh Arquitetos) + Gabriel Castro
Location: Belo Horizonte, Brazil
Completion: 2012
Photography: Gabriel Castro

店面设计：费尔南多·马库兰，玛丽莎·马查多·科埃略（MACh Arquitetos 建筑事务所），加布里埃尔·卡斯特罗
店铺地点：巴西，贝洛奥里藏特市
竣工时间：2012 年
图片摄影：加布里埃尔·卡斯特罗

In the words of the owners, the Emporium Nanak was born of a desire to support a lifestyle seated on the tripod: happy, healthy and holy. The standards used in the drilling of the pannels were extracted from Nanak Emporium's logo, which is in turn referenced in the iconography present at the Golden Temple in Amritsar, a shrine of Sikhism, which is the religion of the owners of the emporium.

A simple system of LED highlights the walls, contributing to product visualization. General lighting is provided by an installation with metal tubes with light bulbs, with controlled intensity.

用店主自己的话来讲，最初经营信徒大市场出于一种对快乐、健康而神圣的生活方式的追求。店面设计中用到的标识取自阿姆利则的金色寺庙中的形象。阿姆利则是印度北部的一个城市，也是店主信仰的锡克教的圣地之一。

简单的 LED 照明系统突出店面墙体，促进产品可视化。其他位置的照明采用金属管配合灯泡，便于控制照明强度。

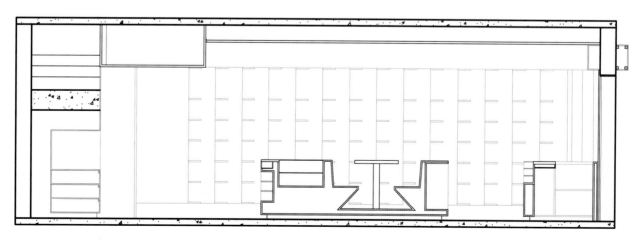

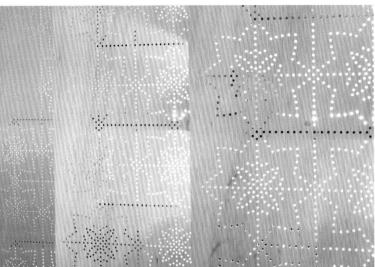

Mango Financial

芒果金融

Design: Bercy Chen Studio LP. Calvin Chen, Thomas Bercy
Location: Austin, Texas. USA
Completion: 2011
Photography: Ryan Michael

店面设计：贝尔西·陈工作室，卡尔文·陈，汤马斯·贝尔西
店铺地点：美国，得克萨斯州，奥斯汀
竣工时间：2011 年
图片摄影：瑞安·迈克尔

Mango Financial is an innovative banking & financial service company using new technology for underserved customers. The concept was to create the financial sector equivalent of the Apple's store.
The design utilizes a secondary clear glass skin which has small frosted imprints of the Mango logos over a gradient of colours from red, orange to yellow. The interiors are customized in a minimalist and functional manner.

芒果金融是一家创新模式的银行和金融咨询服务公司。本项目的目标是打造一个苹果店水准的金融中心。
设计师采用了透明玻璃外墙，上面有芒果金融标识的磨砂小图案以及从红、橙到黄的颜色渐变。室内设计则以简约风格和功能性为主。

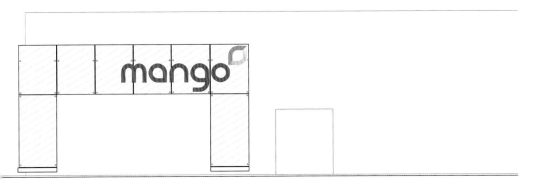

Sugamo Shinkin Bank Ekoda Branch

日本巢鸭信用银行 Ekoda 分行

Design: Emmanuelle Moureaux architecture + design
Location: Tokyo, Japan
Completion: 2012
Photography: Daisuke Shima / Nacasa & Partners Inc.

店面设计：艾曼纽建筑设计公司
店铺地点：日本，东京
竣工时间：2012 年
图片摄影：志满大辅 / Nacasa 摄影合作公司

The site is located in a commercial district with many stores. The site's closeness to the town's activities – also the heavy traffic and narrow sidewalk – inspired the architect to express this proximity in the building by merging the exterior and interior.

The building is offset approximately 2 meters from the property line, and the timber-decked peripheral space is filled with colourful 9 meter-tall sticks. These 29 exterior sticks, reflected on the transparent glazed façade, mix naturally with the 19 interior sticks placed randomly inside the building. This rainbow shower returns colours and some room for playfulness back to the town.

The exterior deck space, interior open space, exterior courtyard, and the interior teller counters compose four layers of spaces. The layers are reflected on the glazing, and, combined with complex shadows, they create depth in the space.

本项目位于繁华的东京商业区内，得天独厚的地理位置伴随着拥挤的交通和狭窄的人行道，这样的景象与特点启发设计师通过店铺内外景观的结合表现这种熙熙攘攘之感。
店面距离街道边缘约 2 米，设计师利用 9 米高的多色细柱状装饰物打造独特的外部空间。这 29 根装饰柱在玻璃外墙上的映像与室内随意摆放的 19 根装饰柱融为一体，为城市增添一抹活泼的色彩。
外部甲板空间、室内的开放空间、室外庭院和室内出纳台构成四个空间层次。这些层次都在玻璃外墙上构成映像，与深邃的阴影一起打造空间的深邃之感。

Money Shop

钱庄

Design: Nicolas Tye Architects
Location: Birmingham, UK
Completion: 2010
Photography: David Helsby

店面设计：尼古拉斯·泰伊建筑师事务所
店铺地点：英国，伯明翰
竣工时间：2010 年
图片摄影：大卫·赫尔斯比

The designers were approached by the Dollar Financial Group to contribute to a re-branding exercise to their portfolio of some 200 'The Money Shop' high street stores. The main objective was to introduce a new prototype that would push the boundaries as set by the current design in order to establish a relationship with a broader demographic by appealing to new customers. The new store concept was introduced in line with the exploration of a new high street location. The minimalist shopfront is in perfect harmony with the building while the all-glass façade with white logo provides clear direction even at night.

设计师受邀为美元金融集团的 200 家钱庄进行店面设计。设计的主要目标是引入一个全新模式，推动业务发展，吸引新客户。店面设计理念同时迎合各个钱庄的具体位置。
该项目店面设计极其简洁，与周围建筑融为一体。全通透的玻璃墙和白色 LOGO 的组合，即使在夜间也为顾客提供了很强的指引作用。

CarClub Firestone

费尔斯通汽车俱乐部

Design: Juan Vazquez
Location: Sao Paulo, Brazil
Completion: 2011
Photography: Juan Vazquez

店面设计：胡安·巴斯克斯
店铺地点：巴西，圣保罗
竣工时间：2011年
图片摄影：胡安·巴斯克斯

The façade, close link between store - customer is the main development. The main concept is to prioritize Carclub Firestone warranty. To this was applied on the façade company brand and complemented with institutional colour auxiliary graph using graphical information regarding the services provided by the customer.

店面外墙是连接店铺和顾客的重要桥梁，本项目的主要设计概念是突出费尔斯通汽车俱乐部的质量担保。这一理念在外墙设计上得以体现，辅以品牌经典配色和平面图案，强化服务内容和特色。

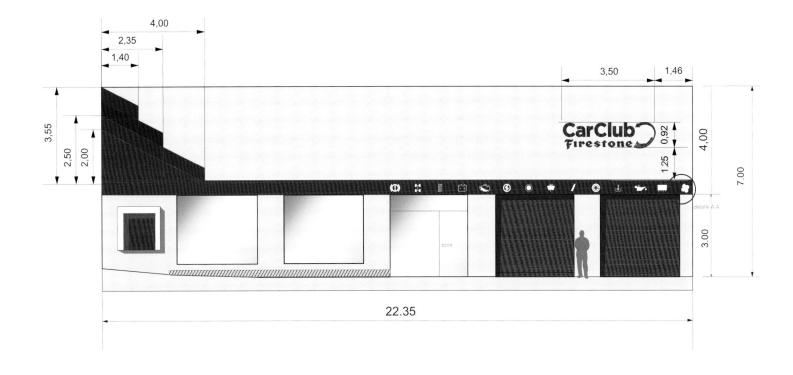

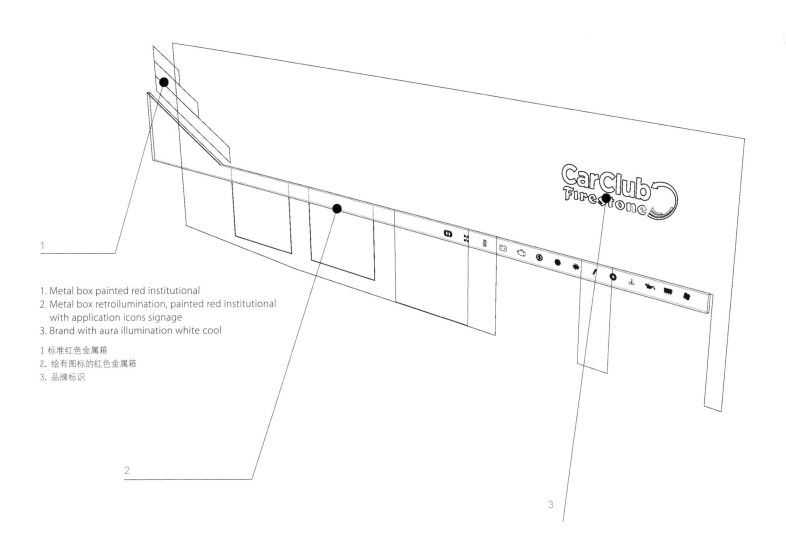

1. Metal box painted red institutional
2. Metal box retroilumination, painted red institutional with application icons signage
3. Brand with aura illumination white cool

1 标准红色金属箱
2. 绘有图标的红色金属箱
3. 品牌标识

Fender Custom Shop Mexico City

墨西哥城芬德吉他行

Design: Arquitectura en Movimiento
Location: Mexico City, Mexico
Completion: 2012
Photography: Cuauhtémoc García

店面设计:"运动的建筑"设计工作室
店铺地点:墨西哥,墨西哥城
竣工时间:2012年
图片摄影:瓜特穆斯·加西亚

Located within the historic center of Mexico City, on the rooftop of one of the most important themed stores of the music scene, the Fender Custom Shop Mexico City is Latin America's first Fender store and follows a different concept from those in other countries.

The basic concept for the development of façade is an irregular polygon, together with the irregular polygon in structure. The inspiration of façade is picked up from the wave of sound and movement of vibration of music. The façade is composed of several sections which slope in different directions and at different angles, generating patterns of natural light. Lighting, a key element in the façade design, was conceived according to use and function. The façade is metal-cladded multipanel, and the sturdiness of this material, its acoustic and thermal properties and its low maintenance make it a suitable choice of finishing.

芬德吉他行位于历史悠久的墨西哥城中心区域,最有影响力的主题乐器商店楼上。这里也是拉美地区首家芬德商铺,采用的是与其他国家和地区的店铺截然不同的店面设计理念。不规则多边形是外墙设计的基本理念,其灵感来自于声波的波形与震动的概念。店面外墙由多个部分组成,这些单元朝不同方向,以不同角度倾斜,形成光影交叠的图案。光照作为外墙设计中的重要元素,在设计中得到了充分的考虑和利用。考虑到坚硬程度,声学和热学性能,以及容易维护等特点,外墙表面选择了金属板材覆盖。

1010 Tsimshatsui Flagship Store

1010 尖沙咀旗舰店

Design: Clifton Leung
Location: Hong Kong, China
Completion: 2012
Photography: Shia Sai Pui

店面设计：梁显智
店铺地点：中国，香港
竣工时间：2012 年
图片摄影：佘世培

Covering some 10,000 sq. ft., 1010 TSIMSHATSUI's signature feature is its 2-storey LED-lit façade. Visibly designed with interconnected yellow light columns, the façade reinforces the distinct 1010 logo colour, and creates a bold statement for the store on the strip, in one of the busiest districts in the city.

LED-lit panels in black frame form a fascinating façade pattern and intensify the 3D effect of the architectural structure. 1010 logo against a white-lit light box defines the space for the two main entrances, while smaller logos on a solid black background are subtly incorporated into the LED columns, to bring forth a visually appealing contrast to maximize the brand presence.

Infusion of modern technology symbolizes the technologically-advanced 1010 brand. Visually captivating imagery are projected on the window façade with 3M translucent film, which resulted in a more lively and vivid exterior to maximize public awareness.

1010 尖沙咀旗舰店占地一万平方英尺（约 929 平方米），其最显著的特点就是两层楼高的 LED 外墙。整个外墙采用黄色照明条相互交错，在强化 1010 馆别致标识色的同时，也在香港最繁忙的地带树立鲜明的店铺形象。

黑框面板组成外墙图案，配合 LED 照明装置，极大地加强了建筑结构的立体效果。1010 馆的大型标识图案以白色灯箱为背景，指示出两个主入口，较小的标识则依托黑色背景，与 LED 光柱巧妙融合，形成鲜明的对比，也使店面的存在感得到进一步强化。

现代科技的使用象征 1010 馆追求技术领先的宗旨，引人入胜的影象投射到贴有 3M 半透明薄膜玻璃的外墙上，使整个建筑外观更加活力四射，有效地提高了公共影响力。

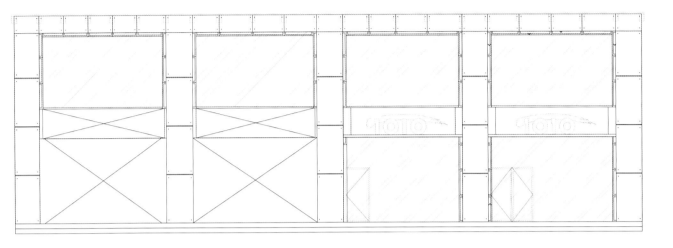

Smartson Store

Smartson 手机商城

Design: Brigada Geber, Damjan Geber, Marina Brletic
Location: Zagreb, Croatia
Completion: 2012
Photography: Domagoj Kunic

店面设计：布里加达·格柏，达米扬·格柏，马里纳·波莱蒂齐
店铺地点：克罗地亚，萨格勒布
竣工时间：2012年
图片摄影：朵马格·库尼奇

Smartson store in the centre of Croatia's capital Zagreb is the biggest specialized smart phone centre in the region. Their friendly and professional approach is shown in the interior and store front design by the usage simple materials like wood and bricks which brings out the contrast with the hi-tech products displayed.

Smartson 手机商城位于克罗地亚首府萨格勒布市中心，是这一地区最大的智能手机经销商。其温馨而不失专业的经营理念通过室内装饰和店面设计体现得淋漓尽致，装饰使用木材和砖石等简洁的天然材料，与陈列、出售的高科技产品形成有趣的反差。

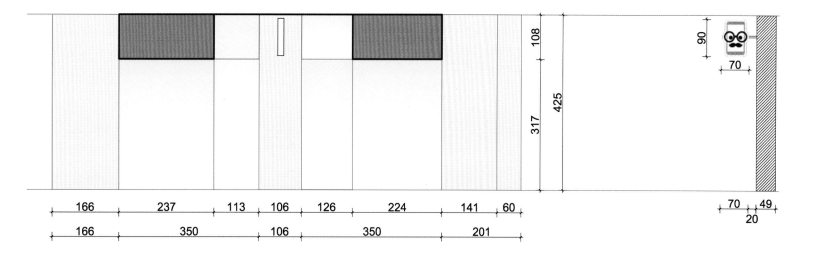

PCCW-HKT Signature Store

香港电讯旗舰店

Design: Clifton Leung /Clifton Leung Design Workshop
Location: Hong Kong, China
Completion: 2013
Photography: Clifton Leung Design Workshop

店面设计：梁显智 / 智设计工房
店铺地点：中国，香港
竣工时间：2013 年
图片摄影：智设计工房

The brand new design of PCCW-HKT flagship store reflects the brand's leading position as Hong Kong's most experienced and all-round telecom service provider, with a century of experience and the largest customer base.

The design concept originates from the historical monument, Arc de Triomphe, Paris which was built as a symbol of victory and power.

The magnificent façade also features a dazzling lightbox with optic fiber graphic, symbolizing the pinnacle of high-speed data communication mediums. Back-lit logo is set against a huge gold panel, injecting a sense of opulence and modernity to the overall design.

Gold is selected as a core colour to accentuate the sense of grandeur and to reflect the leading position of PCCW-HKT. Unique tree column design for mobile display creates a distinctive blend of nature and technology. All these created a distinctive identity for the PCCW-HKT new image store rejuvenation.

香港电讯的全新旗舰店设计充分反映品牌在行业中的领先地位。香港电讯的百年历史和庞大客户基础使其稳居香港最具经验的全方位电信服务运营商之席。

旗舰店设计理念源于象征胜利和力量的巴黎凯旋门。

富丽堂皇的店面外观采用炫目的光纤特效灯箱，代表高速的数据通信媒体。背光标识以巨大的金色板材为背景，为整体设计增添华丽与现代之感。

作为旗舰店设计的主色调，金色突出店面的宏伟气氛，强化品牌的领先形象。独特的树式结构用于移动展示，将自然与科技结合在一起，让人印象深刻。品牌蓬勃的生命力在全新的品牌形象中呼之欲出。

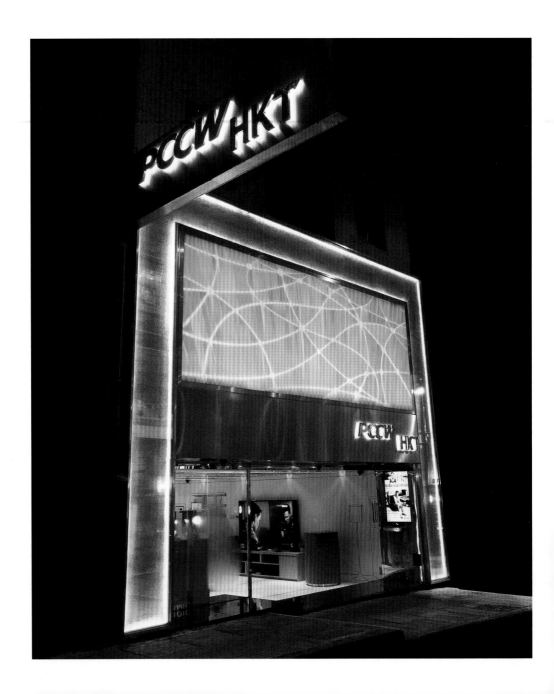

1. 100mmTHK. Light box
2. Matt gold finish
3. Embedded TV
4. Clear glass

1. 100毫米厚灯箱
2. 金色亚光表面
3. 嵌入式电视机
4. 白玻璃

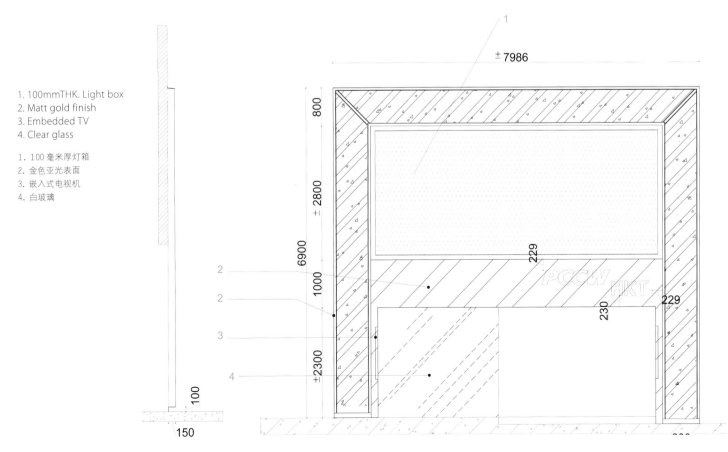

Venus Jewellery

维纳斯珠宝店

Design: Mamen Domingo, Ernest Ferré
Location: Barcelona, Spain
Completion: 2010
Photography: Nagore Linares

店面设计：玛门·多明戈，欧内斯特·费雷
店铺地点：西班牙，巴塞罗那
竣工时间：2010年
图片摄影：纳戈雷·利纳雷斯

Venus Jewellery is located in the city center of a location near Barcelona. The plot has always shared living and commercial uses. The project proposes a radical change, so it wants all the volume to be understood as a unit.

The project begins with a full demolition of the inside. The entire ground floor ceiling and the last part of the basement are pulled down. The new frameworks let a new long double height in between ground and first floor and another one towards the basement. The consecutive double heights join the last three floors in a unique continuous space. These heights are the main character of the space, so they are turned into a big shop window, a big jewellery container with elongated showcases, which begin at the outside and are extended all the length and the height of the inside. Jewels are the organizing element of the project. At the façade, the domestic image is converted into a more actual commercial one, where the importance of the indoor showcases continues. The empty of the ground floor and the big hole in the first floor are framed by the sides, where a couple of showcases will always be visible, as if they had run away of the inside exhibitions to be shown to the city.

维纳斯珠宝店坐落在巴塞罗那市中心一处商住两用的建筑之内。工程要求对店面进行大幅度的改造，将整个建筑凝聚成一个整体。

店面内部被全部拆除，整个一楼的顶棚和地下室的残余结构被拆掉。新方案里，一楼分别与二楼和地下室相连，构成双层高度的室内空间。这样的设计让这三层楼组合成一个独特的连续空间，宽敞的室内空间配合高大的橱窗和珠宝箱元素，与贯穿其间的珠宝形成了让人过目难忘的店面展示。

外墙设计延续了内部的元素和风格。陈列柜将宽敞的一楼空间与二楼的开口连接在一起，好像要带着店内珠宝飞离店面，具有很强的视觉吸引力。

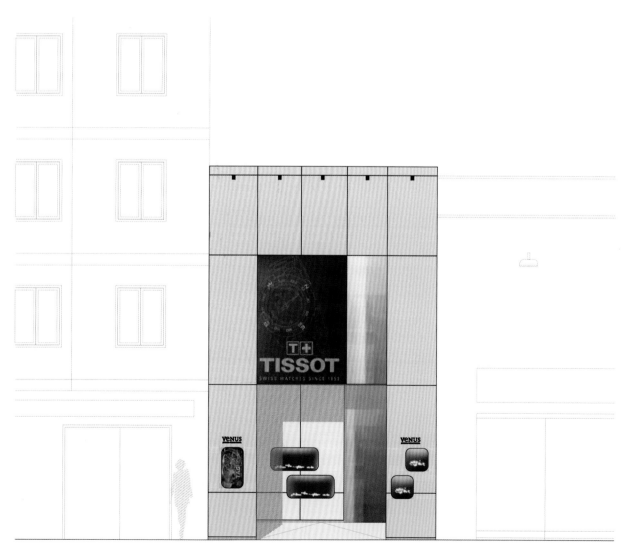

Centro Felicita

Centro Felicita 珠宝店

Design: Kiyoshi Miyagawa
Location: Sendai, Japan
Completion: 2011
Photography: SURE SiGN

店面设计：宫川清志
店铺地点：日本，仙台
竣工时间：2011 年
图片摄影：Sure Sign 摄影公司

This plan is for a bridal jewellery store which locates at a shopping street in downtown Sendai, Miyagi. The designers did total design planning as well as interior, graphic and website. Designers chose white as the main colour of the shopfront for its pureness and holiness. Large glass façade at the entrance and illuminated logo are employed to present a distinctive brand image.

The plan for the façade inspires intellectual curiosity, and what we do is just showing the first floor entrance from the store façade for making customers want to go to the second floor. And to make people feel easy to enter the store, designers put a blank space by putting a space among main items near the entrance.

这是为一家坐落于仙台宫城县商业区的婚嫁珠宝店进行的店面改造。设计师受邀进行了包括室内、平面和网站在内的整体规划设计。外部墙面选用纯洁而低调的白色，同时入口处大量使用了全通透的玻璃，搭配以该品牌灯光效果的LOGO，树立起鲜明的品牌形象。

外墙设计努力激发人们的好奇心，店面的第一道入口吸引人们进入店内探索第二道门。设计师特别在靠近入口处设置了留白，加强进入的方便之感，吸引潜在顾客前来一探究竟。

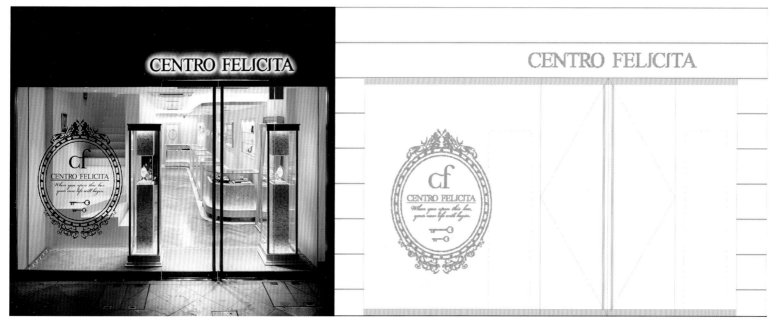

Rosa Jewellery Shop

罗莎珠宝店

Design: Puntidifuga
Location: Mondovi, Italy
Completion: 2011
Photography: Paolo Strobino

店面设计：Puntidifuga 设计公司
店铺地点：意大利，蒙多维
竣工时间：2011 年
图片摄影：保罗·斯多比诺

The Rosa Jewellery designed by Paolo Strobino director of Puntidifuga, is the refurbishment of an existing shop managed by the same family for more than 30 years in the heart of the old town of Mondovì (Italy).

The external façade is a brilliant restyling of the old design survived from the past. It was mandatory, due to budget reasons, to keep all the structure, window glass and security shutters of the existing façade. The idea was to convey a strong impact reducing the area of transparent glass, and giving to the whole façade the same colour. Therefore 5mm marine plywood panels with scratched resin have been glued to the existing window glass in order to reduce and focus the display area. All the original decorations and gold mouldings have been painted with the same dark grey colour of the blind parts. Bronzed metal with brushed finish has been used for the trims around the displays, the door frame and the signage plate. The result is an original and modern design which reveals traditional and more classic atmosphere and details.

罗莎珠宝店的店面改造项目由 Puntidifuga 设计公司的保罗·斯多比诺主持设计，主要内容是对原有店面的翻新。这家首饰店位于意大利老城蒙多维，由同一家族经营了30多年。外墙设计是在原有基础上进行的出色改造。由于预算限制，需要保留所有原来的建筑结构、玻璃橱窗和安全门。这次店面设计的主要内容是减少玻璃窗的面积，突出展示的重点，统一整个店面的外观颜色。因此设计师在原有的玻璃窗上黏合了有树脂刻痕的5毫米胶合板材，以便减少展示空间，加强展示效果。所有原来的装饰物和金色造型都刷上了与百叶窗部分相同的深灰色。展示空间、门框和标牌的边框使用了刷面处理的青铜色金属。最终的设计风格独特，具有现代感，同时又保留了传统的经典细节。

1. Existing mouldings to be painted as per resin colour
2. Scratched resin applied on wooden panel in place of existing mirror
3. Cut out plexi logo
4. Vertical metal posts to be painted as per resin colour
5. New door handle covered by resin
6. Existing stone skirting to be covered by resin
7. New metal profile to match exiting vertical profiles
8. Showcase metal frame thick. 3mm, bronzed brass, brush finish with glossy transparent coating
9. Scratched resin applied on wooden panel glued on existing window glass

1．原有造型用树脂涂料上色
2．木板上涂刷树脂涂料替代原有镜面
3．店铺标识
4．垂直金属柱用树脂涂料上色
5．涂刷树脂的新门把手
6．原有石材踢脚板用树脂涂料涂刷
7．新的金属框架匹配原有框架
8．3毫米厚金属展示架，透明光面涂料
9．粘在原有橱窗上的木板上涂刷树脂涂料

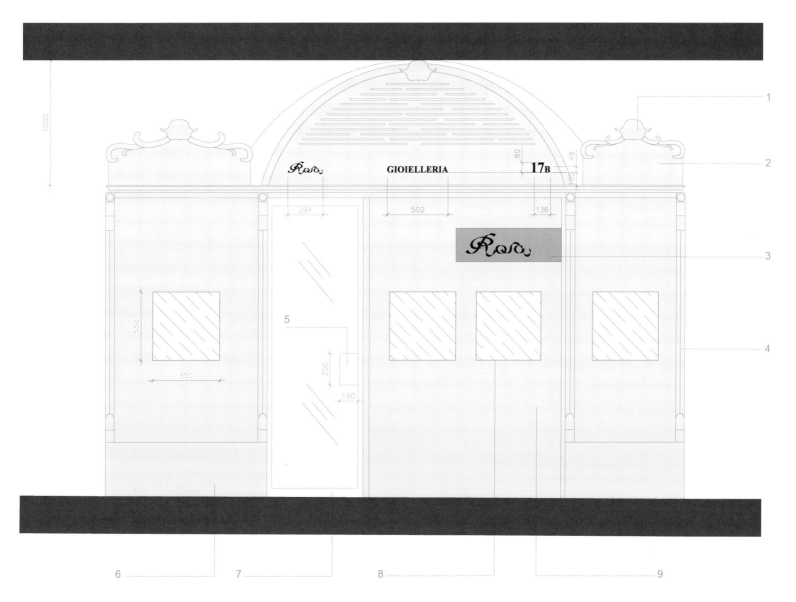

Burma Jewellery Boutique, Rue de la Paix

缅甸珠宝精品店

Design: Anne-Cécile Comar, Philippe Croisier, Stéphane Pertusier /Atelier du Pont
Location: Paris, France
Completion: 2013
Photography: Philippe Garcia

店面设计：安妮·塞西尔·科马尔，菲利普·克罗瓦吉耶，史蒂法纳·柏图塞吉耶 / 杜邦工作室
店铺地点：法国，巴黎
竣工时间：2013 年
图片摄影：菲利普·加西亚

The front of the Burma boutique in Rue de la Paix is decked out in black and glass. The exterior and interior spaces attract and interact via a façade that takes its inspiration from the portrait gallery. The jewellery in the shop window is exhibited in hanging elliptical display cases that almost seem to float in the air; the back of the cases is reminiscent of a hand-held mirror. This display provides a bold and unusual introduction to the store.

位于巴黎宁静街上的这家缅甸珠宝精品店以黑色和玻璃材质为主。外墙是店内外空间互动的媒介，其设计灵感来自肖像馆设计。橱窗内悬挂起来的椭圆形展示单元似乎飘浮在空气中，展示单元的背面让人联想起精致的手持镜子。这种大胆的珠宝展示方式，令人印象深刻。

Swarovski

施华洛世奇阿里坎特店

Design: Manuel García Estudio
Project management: Manuel García Estudio
Location: Alicante, Spain
Completion: 2012
Photography: ClaroOscuro Fotografía

店面设计：曼努尔·加西亚工作室
项目管理：曼努尔·加西亚工作室
店铺地点：西班牙，阿里坎特
竣工时间：2012年
图片摄影：克拉罗·奥斯库罗摄影

This Project shows the Swarovski Shops new image with a totally renewed design, which is a result of a constant searching for innovation, modernity and perfection. The concept receives the name of 'Crystal Forest' and its aim is to highlight the endless possibilities of glass, as well as to show the affinity with the natural world.

The situation of the shop in a corner together with its two large glazed fronts offers an open exhibition concept, transmitting outside light and freshness and inviting you to visit the interior.

The shop window is defined by a totally transparent skin which frames the space, and without offering visual barriers, turns the whole surface of the shop into the own window display itself. Nevertheless, in the foreground, a small selection of the product is showed by using two display cabinets of different nature, which are customizable according to the new collections.

On the broad windows features a bright and textured strip that frames the sign, where the surfaces come alive with a reflective prisms of different lengths and volumes. These multilayered faceted elements create visual depth and give off a rhythm in line with rules and customs of nature.

经过不懈的探索和努力，重新设计后的施华洛世奇店面呈现出全新的面貌。这个名为"水晶森林"的设计理念结合了现代风貌和精益求精的精神，突出玻璃制品的无尽可能，展示自然的无穷魅力。

店面位于角落，两扇高大的玻璃外墙便于商品展示，将商品的光泽和精致透过玻璃展现给消费者，激发消费欲望。

橱窗在这里被定义为完全透明的无障碍视觉元素，它限定了店面的界限，同时将整个店面转化为一个大橱窗。前景中是根据新主题定制的不同材质的两个展示柜，陈列着一小组新产品。

鲜亮的条状织物围住店面标识，配合不同大小的棱镜，使宽阔的橱窗焕发出生机与活力。这些多层面元素构成视觉的纵深感，象征自然世界中的规则和韵律。

Zlatarna Celje Jewellery Flagship Store

Zlatarna Celje 珠宝旗舰店

Design: OFIS ARCHITECTS
Location: Maribor, Slovenia
Completion: 2011
Photography: Tomaz Gregoric, Jan Celeda, Giulio Margheri

店面设计：OFIS 建筑师事务所
店铺地点：斯洛文尼亚，马里博尔
竣工时间：2011 年
图片摄影：塔马兹·格雷戈里克，让·谢利达，朱利奥·马格里

There are three main store typologies; high street flagship stores, shopping center shops and gold investment centers. The design is based around a core concept for the Zlatarna Celje brand that its customers will identify in which ever of the three typologies because they each use a variation of this principal idea.

The project concept derives from the idea of a bank safety deposit box, using the simple idea of a strong, protective box safeguarding valuable jewellery. A box module is repeated through out the store to create a sense of continuity and repetition, using the same material and module.

The investment center, however, is more reminiscent of a steel bank vault in feel and appearance. The safety deposit boxes create a slightly more formal feeling since business proceedings take place within this type of store.

The simple use of two different materials for the two store typologies create an obvious visual difference between them for the customer to distinguish.

Zlatarna Celje 珠宝品牌拥有三种类型的店铺：高街旗舰店、购物中心店和黄金投资中心。本项目的设计围绕 Zlatarna Celje 珠宝品牌的核心理念，无论是哪种类型的店面，顾客都可以轻松辨识。

本项目的设计灵感源自银行保险箱的概念，使用结实、安全的箱子保护珍贵的珠宝。保险箱的形象通过相同的材料和形式在店内反复出现，呈现连续、统一的氛围。

投资中心的设计有所不同，外观看起来像一个钢铁制成的金库。投资中心是进行商业交易的场所，在这里使用保险箱的形象有助于打造更为正式的氛围。

两种类型的店面使用不同的材料，形成强烈的视觉反差，利于顾客进行辨识。

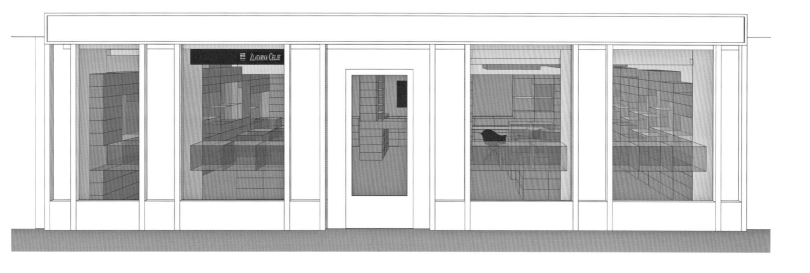

Herman Miller Shop-in-Shop - XTRA

赫尔曼·米勒新加坡店中店

Design: P.A.C Pte Ltd
Location: Singapore
Completion: 2012
Photography: Stacey Peh

店面设计：P.A.C 设计公司
店铺地点：新加坡
竣工时间：2012 年
图片摄影：史黛丝·佩赫

The design inspiration comes from Herman Miller's core products: Work Chairs. Herman Miller's extensive range of work chairs produced over the years embodies their philosophy, culture and aspirations; a desire to constantly innovate, integrate technology with design and produce high quality, purposeful, human-centered products.

The Herman Miller store is expressed as a porous plywood envelope forming an independent entity that plugs into the shop space of Xtra. This lightweight envelope allows visual links yet maintains the identity of both stores.

Just as the Herman Miller work chairs are designed to adapt to our postures and movements, the Herman Miller skin is moulded to adapt to existing structures and customer movement patterns. The result is a skin informed by Herman Miller's philosophy that creates an identity inspired by their core body of work. The parametric surface modulates light and views into a flexible, open space. The plywood material forms a warm and casual ambience while its structure and assemblages expresses the dynamic process that combines technology with design.

设计灵感来自赫尔曼·米勒品牌的核心产品：工作椅。多年以来，赫尔曼·米勒品牌出产了种类繁多的工作椅，充分体现了品牌的理念、文化和远大抱负；对不断创新的渴望，实现科技与设计的整合，生产高质量、意义深远、以人为本的产品。

赫尔曼·米勒的这个店面采用多孔胶合板外墙造型，并延伸入店内空间。轻质外墙材料在视觉上起到统一、连接的作用。

正如赫尔曼·米勒工作椅是为了适应人们的姿势和动作而设计的，这项店面设计也针对原有建筑及人群流动模式进行了相应的调整。外墙表面对光线有调节作用，视角灵活开阔。胶合板营造出温馨休闲的气氛，其结构和组合形式则展现出动态之美。

BASE camp

BASE camp 数字生活休闲店

Design: NEST ONE
Location: Berlin, Germany
Completion: 2012
Photography: NEST ONE

店面设计：NEST ONE 设计公司
店铺地点：德国，柏林
竣工时间：2012 年
图片摄影：NEST ONE 设计公司

Right in the heart of Berlin opens the new 330m² shop combining selling space, café, co-working space, training room and event location. The store design with minimalist style, black door frames, window frames with glass, mix and match black and white logo, to bring out the best in each other.

这家新店铺位于柏林市中心，总面积330平方米，集零售空间、咖啡厅、工作区域、锻炼中心和活动场地于一身。店面设计以简约风格为主，融合了黑色门框、窗框，玻璃橱窗，混搭风格的黑白标识，呈现个性的店面形象。

Novo Ambiente Design Store

"新环境"家具店

Design: Ivan Rezende
Location: Rio de Janeiro, Brazil
Completion: 2011
Photography: Ivan Rezende

店面设计：伊万·雷森德
店铺地点：巴西，里约热内卢
竣工时间：2011 年
图片摄影：伊万·雷森德

The original building of 3 floors which is situated in the traditional Rua Redentor was completely uncharacteristic in its architecture, and it was then entirely revitalized, gaining a new façade.

This concept of façade elevates and enhances the products displayed, creates a volumetric difference from the planes and lines and points bringing greater visibility to the designer pieces. Façade lines relate to the lines forming the inner planes.

项目所在的三层建筑原本并不起眼，精心的设计和改造使它重新获得了活力。

外墙设计理念的提升加强了展示商品的视觉影响力，在点、面、线的层面上营造出焕然一新的店面形象，使店铺焕发生机。

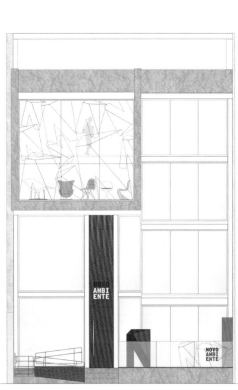

Black and White Gobbi Novelle Store

戈比·努维勒家具店

Design: Henrique Steyer, Felipe Rijo /ALBUS Design
Location: Porto Alegre, Brazil
Completion: 2012
Photography: ALBUS Design

店面设计：恩里克·斯泰尔，费利佩·赖伊 /
ALBUS Design 设计公司
店铺地点：巴西，阿雷格里港
竣工时间：2012 年
图片摄影：ALBUS Design 设计公司

Classic are always in fashion, and the colours black and white dominate the winter/fall decoration of façade for the store Gobbi Novelle in Porto Alegre, Brazil. The new façade was created by architect Henrique Steyer from ALBUS Design, who used as inspiration and basis for the concept the quote by Aristotle Onassis: 'I dye my hair white for business meetings, and black for romantic outings.' The 'Black and White' concept begins on the entryway stairs, half-painted on each colour. The façade follows the same paint scheme.

经典永不过时，而巴西阿雷格里港的这家店铺在秋冬季的店面设计中采用了黑与白的经典配色。店面外墙由 ALBUS Design 设计公司的设计师恩里克·斯泰尔打造，他的设计灵感源自亚里士多德·奥纳西斯的名言："商务会见时我将头发染成白色，浪漫约会时我将头发染成黑色。""黑与白"的设计理念从半黑半白的入口楼梯开始，外墙也是如此，相互呼应。

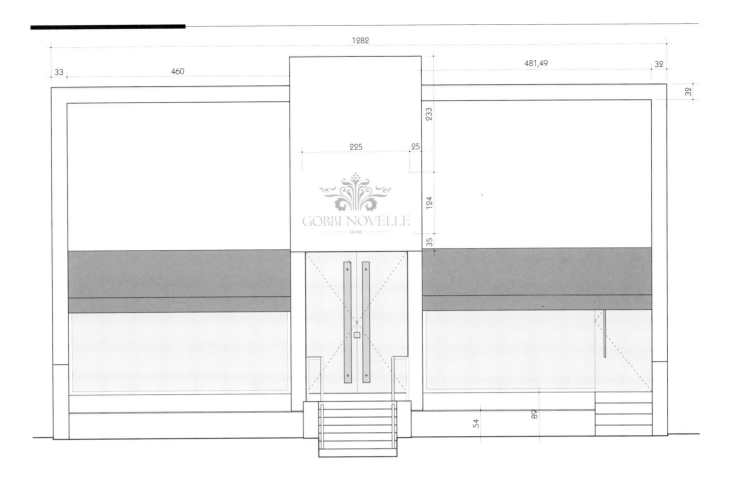

Mobica Prado Norte

莫比卡家具店

Design: ARCO Arquitectura Contemporánea
Location: Mexico City, Mexico
Completion: 2009
Photography: José Lew

店面设计：ARCO 当代建筑设计公司
店铺地点：墨西哥，墨西哥城
竣工时间：2009 年
图片摄影：何塞·卢

This new branch Mobica- located at Prado Norte, in the Lomas de Chapultepec- is designed as a simple space, sober and Vanguard, but does not lose the warmth that is required in a store of furniture and home accessories. Outside the store has a great impact, since its purpose is to highlight the environment and invite the clients to discover inside.

The store has two levels. The main façade placed a window to the level of access that leads directly to the main exhibition and sales area. In the cabinet exterior lighting has a system that makes it stand out and give the impression of floating. At the top of the façade was installed a modular system that gives a nice motion effect.

莫比卡家具店的这家新店位于墨西哥城的北普拉多，店面设计风格简洁、前卫，却又不失家居装饰店的温馨气氛。店面外观运用了明亮的色彩作为品牌名称，能够有效地吸引顾客进店探索感兴趣的产品。

商店占据两层楼。橱窗与大门处于水平位置，后者直接通向主展厅和销售区域。室外照明系统营造出飘浮的感觉。外墙顶端的模块系统也具有动态美感。

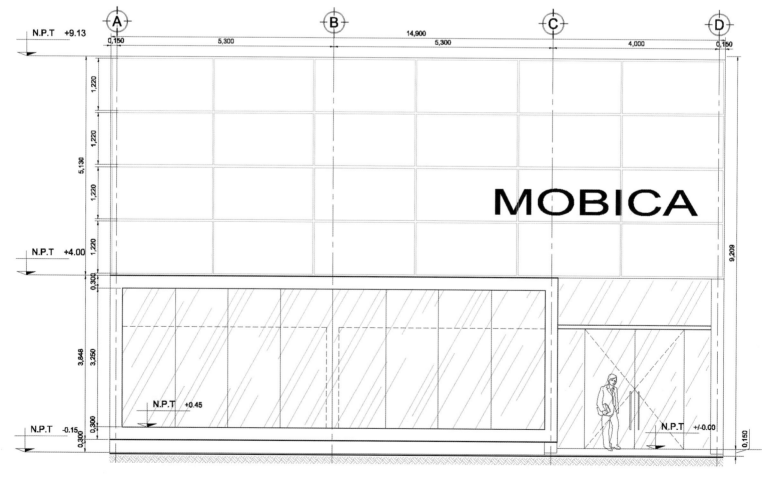

Skitsch Store Milan

Skitsch 家具米兰店

Design: Blast Architects
Location: Milan, Italy
Completion: 2009
Photography: Blast Architects

店面设计：Blast Architects 建筑师事务所
店铺地点：意大利，米兰
竣工时间：2009 年
图片摄影：Blast Architects 建筑师事务所

Blast Architetti designed the project for the first Skitsch store, a new Italian design brand. The six-hundred-square-meter thirteen-window store is on the corner of Via Monte di Pietà and Via Fratelli Gabba in the heart of downtown Milan. Movement, interaction, rhythm, and a multi-sensory approach were the key words that inspired choices of façade. This solution of façade offers a direct contact – a relationship – with the various objects and it also provides a view of the entire store through the windows opening onto the street. A search for simplicity in façade is expressed through natural minimalism: the building's structure. The simplicity of the pure white walls, glass, and bronze (another reference to the historic building) underlines the formal rigour of this design project. The only concession to colour is the institutional blue of Skitsch.

设计师为意大利的全新家具品牌 Skitsch 家具设计了第一个店面。这处 600 平方米的店面有 13 个窗户，坐落在米兰市中心最繁华的区域。

动态、互动、节奏和多重感官是店面设计中的几个关键词。外墙设计方案通过橱窗将店内与室外连接在一起。设计师对简约风格的追求表现在外墙自然简单的表现方法上。纯白的墙壁、玻璃和青铜（代表建筑的历史）体现了工程的正式与严谨。设计中做出的妥协是使用到了品牌本身的独特蓝色色调。

PART 8 Works-Furniture

Skitsch Store London

Skitsch 家居店

Design: Blast Architects
Location: London, UK
Completion: 2010
Photography: Blast Architects

店面设计：Blast Architects 建筑师事务所
店铺地点：英国，伦敦
竣工时间：2010 年
图片摄影：Blast Architects 建筑师事务所

For the new store in London, Blast Architects has used natural materials and enhanced the façade in order to create a strong bond between rough concrete walls and glass curtain wall. On the glass wall it says, 'We wish you a merry design and a happy new Skitsch.' The concept of façade is in accordance with the words on the glass wall. Tactility and simplicity transmit the minimalism to happiness, and the transparency and frame in façade are such as the image of reminiscence inset in the frame.

设计师选用了天然建筑材料，强化店面外墙，在粗糙的混凝土墙和玻璃幕墙之间构建紧密的关联。玻璃幕墙上的文字写道："欢迎选择 Skitsch 家居产品，将设计带回家。"外墙设计理念与之紧密契合。感性、简洁的设计传递着简约而又幸福的生活氛围，店面外墙的透明度和框架设计精致巧妙。

Fiori Flower Boutique

菲奥里精品花店

Design: Belenko Design
Location: Kiev, Ukraine
Completion: 2011
Photography: Andrey Avdeenko

店面设计：别连科建筑设计公司
店铺地点：乌克兰，基辅
竣工时间：2011 年
图片摄影：安德烈·阿夫迪恩科

Fiori Flower Boutique is located in downtown of Kyiv. The elevation looks very cosy because of using wood in a part of it. One comes outside from the interior through the big storefront, and wooden entrance doors and a portal that look like a part of the interior - it has the same finishing as the interior, which is covered with wood. Traditional balusters and columns are painted in matte grey. Bright blue sign and big black sconces underline elegant eclectic façade.

菲奥里精品花店位于基辅市中心。立面造型中的木材元素给人以休闲舒适的感受。店面大门与室内环境使用了相同的木料表面，形成统一连贯的视觉感受。传统造型的栏杆和柱子粉刷成亚光灰色，亮蓝的标识和黑色烛台突出店面优雅的外观。

INDEX 索引

3GATTI

A
A-cero Joaquin Torres & Rafael
act_romegialli
ALBUS Design
Ali Alavi
Alice Tepedino
Ana Luiza Neri
Angela Maria Romegialli
ARCO Arquitectura Contemporánea
Arquitectura en Movimiento
Arquitectura y Diseño
Atelier du Pont
Ayla Carvalhaes

B
Baar-Baarenfels Architekten
be.bo.
Belenko Design
Bel Lobo
Bercy Chen Studio LP.
Blast Architects
Bob Neri
Bonetti Kozerski Studio
Brigada
BXH Arquitectura

C
Campaign
Carla Dutra
Carol Kaphan Zullo
Checkland Kindleysides
Christopher Ward Studio
churtichaga+quadra-salcedo architects
Clarisse Palmeira
Claudiu Toma
Clifton Leung Design Workshop
CRIO arquiteturas
CuldeSac™

D
Damjan Geber
dARCHstudio
Dear Design
Droguett A&A Ltda.
Duccio Grassi Architects

E
Electric Dreams
emmanuelle moureaux
Erika Gaggia architects
Ernest Ferré
estudio 30 51
Estudio Vitale
everedge.inc

F
Fausto Castañeda Castagnino
Fernanda Carvalho
Fernanda Mota
Fernando Maculan

G
Gabriel Castro
Gianmatteo Romegialli
Golucci International Design
gpstudio
Green Room Design Team
Gruppo C14 srl
Guan Design
Gundry & Ducker Architecture

H
Hangar Design Group
Hassan Hamdy Architects
hwayon

I
Iosa Ghini Associati
Ivan Rezende

J
J.C. Architecture
Jenner Studio
Juan Vazquez

K
Kamitopen Architecture-Design Office co.,ltd.
KC design studio
Keiichi Hayashi
Kengo Kuma & Associates
Key Operation Inc./Architects
Kiyoshi Miyagawa

L
lajer & franz studio
LaurentT Deroo Architecte
Les Eerkes

Lisiane Scardoelli
Iosa Ghini Associati
LRS Architects, Inc.
Luciana Carvalho

M
M4
MAM architecture
Mamen Domingo
Manuel García Estudio
Marcia Arteta Aspinwall
María Velásquez
Mariana Travassos
Mário Wilson Costa Filho
Mariza Machado Coelho
Mathias Klotz arquitectos
Mizzi Studios
Monovolume Architecture + Design

N
Nabito Architects
NC Design & Architecture Ltd
Nest One
Nicolas Tye Architects
Norbert Ianko
Nota Design Group

O
ODVO arquitetura e urbanismo
Oficina Mutante
OFIS Architects
OHLAB
Olson Kundig Architects

P
P.A.C Pte Ltd
Patricia Fontaine
Pierluigi Piu
PIUARCH
plajer & franz studio
PLANNING ES
Plexo Design Lab
Plotcreative Interior Design Limited
PROCESS5 DESIGN
Puntidifuga

R
Renato Diniz
Rodriguez studio architecture p.c
Rolf Ockert Design
Ruth Tjitra

S
SAKO Architects
Sofia Mora
Specialnormal Inc.
SPRS Arquitectura
Stefano Tordiglione Design Ltd
Stone dsgns
Studio Cinque
Studio Gascoigne
studio KMJ
SWeeT CO.,ltd
switch-lab inc.

T
T2.a Architects

Teun Fleskens
Tom Kundig
Torafu Architects
Arquitectura y diseño

U
Urbantainer Co., Ltd.
UXUS

V
Vanja Ilić Architecture
VAUMM architecture & urban planning
Verdego
Vincent Choi

W
Wesley Liu

Y
YOD Design Lab
Younghan Chung
Yukio Hashimoto

Z
Zeno Ardelean

图书在版编目（CIP）数据

商业店面设计 /（意）陶迪利诺编；张晨译. -- 沈阳：辽宁科学技术出版社，2015.3
 ISBN 978-7-5381-8914-8

Ⅰ. ①商… Ⅱ. ①陶… ②张… Ⅲ. ①商店—室内装饰设计 Ⅳ. ①TU247.2

中国版本图书馆CIP数据核字(2014)第268789号

出版发行：辽宁科学技术出版社
　　（地址：沈阳市和平区十一纬路29号　邮编：110003）
印　刷　者：利丰雅高印刷（深圳）有限公司
经　销　者：各地新华书店
幅面尺寸：230mm×290mm
印　张：40
插　页：4
字　数：50千字
出版时间：2015 年 3 月第 1 版
印刷时间：2015 年 3 月第 1 次印刷
责任编辑：刘翰林　于　芳
封面设计：周　洁
版式设计：周　洁
责任校对：周　文
书　　号：ISBN 978-7-5381-8914-8
定　　价：358.00元

联系电话：024-23284360
邮购热线：024-23284502
E-mail: lnkjc@126.com
http://www.lnkj.com.cn
本书网址：www.lnkj.cn/uri.sh/8914